To Improve Human Health

A HISTORY
OF THE
INSTITUTE
OF
MEDICINE

Edward D. Berkowitz

NATIONAL ACADEMY PRESS
Washington, D.C. 1998

NATIONAL ACADEMY PRESS • 2101 Constitution Avenue, N.W. • Washington, DC 20418

Additional copies of *To Improve Human Health: A History of the Institute of Medicine,* are available for sale from the National Academy Press, Box 285, 2101 Constitution Avenue, N.W., Washington, DC 20055. Call (800) 624-6242 or (202) 334-3313 (in the Washington metropolitan area), or visit the NAP's on-line bookstore at **www.nap.edu.**

For more information about the Institute of Medicine, visit the IOM home page at **www2.nas.edu/iom.**

International Standard Book No. 0-309-06188-1

Printed in the United States of America

Preface

In this first formal history of the Institute of Medicine (IOM), Professor Edward Berkowitz describes many of the important events and individuals associated with this institution. Many more decades may have to pass before we can fully appreciate or understand the significance of the Institute's work in the larger context of American life. However, the opportunity for Dr. Berkowitz to interview so many of the people who were part of the Institute's creation including each of its presidents warranted a scholarly appraisal at this time.

Memories alone are not necessarily history. Many of those who have been active in the Institute during its short life may have different perspectives about the people and events described in this book. My own view, however, is that Dr. Berkowitz has done an admirable job of capturing both the spirit and substance of the events of the past three decades.*

As I read this history, I was struck by two characteristics of the Institute. Despite many ups and downs, successes and near misses, intellectual disagreements, and financial concerns, certain core values have persisted, as has a pervasive sense of optimism that these values can contribute to better health for all people, nationally and internationally. Those core values and that optimism are clearly reflected in this text.

From its creation, IOM has been committed to:

- knowledge and policies that improve health for present and future generations;
  - objective, impartial evidence-based decision-making;
  - anticipating and confronting difficult issues;

---

*People with additional information, documents, pictures, or perspectives relevant to the history of the Institute are encouraged to send them to the Institute's Reports and Information Office, where they will be collected and transmitted to our archives to benefit future historians.

- interdisciplinary, multifaceted perspectives;
- self-assessment and continuous improvement of its operations; and
- the joining of scientific and humanistic values.*

Many of the issues that seemed to have imminent solutions when the Institute was founded in 1970 continue to challenge us. There still remain uncompleted agendas for the Institute to tackle, such as access to quality health care for all Americans, the strength of the public health infrastructure, the availability of appropriate care for children and the elderly, America's role in international health, issues of nutrition and food safety, the appropriate training and use of the health care work force, and adequate support for the development of new knowledge in these—and other—areas. As this history documents, the Institute has made important contributions in most of these areas and significant progress in others. With a recollection of the past, we rededicate ourselves to making further progress on these issues in the future.

Acknowledgments

The Institute has a wonderful tradition that members share in the work instead of delegating it. That tradition is exemplified by the diligent efforts of the IOM History Oversight Committee in guiding the progress of this project, for which I am deeply grateful. They worked with and advised Dr. Berkowitz as he researched and wrote the manuscript, then provided critical comments and suggestions about the draft and its revision. The committee members were

Adam Yarmolinsky, LL.B. (*Chair*), Regents Professor of Public Policy in the University of Maryland System;
Gert H. Brieger, M.D., Ph.D., William H. Welch Professor and chairman of the Department of the History of Science, Medicine, and Technology, Johns Hopkins University School of Medicine;
Roger J. Bulger, M.D., president, Association of Academic Health Centers, Washington, D.C.; and
Elaine L. Larson, R.N., Ph.D., professor in pharmaceutical and therapeutic research, Columbia University School of Nursing.

---

*From the 1997 Institute of Medicine strategic plan, available on line at: **www2.nas.edu/iom/stratplan.pdf.**

I also appreciate the assistance provided by the individuals who reviewed the manuscript:

Enriqueta C. Bond, Ph.D., president, Burroughs Wellcome Fund, Durham, N.C.;

Barbara J. Culliton, correspondent-at-large, *Nature Medicine,* Washington, D.C.;

Donald S. Fredrickson, M.D., scholar, National Library of Medicine, National Institutes of Health, Bethesda, Md.;

Robert J. Glaser, M.D., independent consultant, Menlo Park, Calif.;

David A. Hamburg, M.D., President Emeritus, Carnegie Corporation of New York, New York City;

John R. Hogness, M.D., President Emeritus, University of Washington;

June E. Osborn, M.D., president, the Josiah Macy, Jr., Foundation, New York City;

Frederick C. Robbins, M.D., University Professor Emeritus, Department of Epidemiology and Biostatistics, Case Western Reserve University;

Samuel O. Thier, M.D., president and chief executive officer, Partners HealthCare System, Inc., Boston; and

Karl Yordy, health policy consultant, Tucson, Arizona.

Other persons—some of whom are already acknowledged here—gave generously of their time and knowledge through interviews conducted by Dr. Berkowitz. They are Stuart H. Altman, Ph.D.; Roger J. Bulger, M.D.; James D. Ebert; Rashi Fein, Ph.D.; Donald S. Fredrickson, M.D.; Robert J. Glaser, M.D.; David A. Hamburg, M.D.; Ruth S. Hanft, Ph.D.; Karen Hein, M.D.; John R. Hogness, M.D.; Irving M. London, M.D.; David Mechanic, Ph.D.; Henry W. Riecken, Ph.D.; Frederick C. Robbins, M.D.; Samuel O. Thier, M.D.; and Adam Yarmolinsky, LL.B.

The long association of each of the abovenamed individuals with the Institute, and their understanding of its triumphs, as well as of its trials and tribulations, make them an invaluable source of knowledge and guidance in the completion of this project

I am also grateful for the help provided throughout the project by a number of IOM staff, including Conrad Baugh, Mike Edington, Karen Hein, Sandra Matthews, Janet Stoll, and Jana Surdi.

Further, I appreciate the generous grant provided by the Burroughs Wellcome Fund, which has partially supported the publication of this book.

Most of all, I thank Dr. Edward Berkowitz for the splendid job he has done in writing this history. Although the IOM produces many, many reports each year, this publication is unique: It is a book *about* the Institute, not *from* the Institute. This is Dr. Berkowitz's contribution to the institution and to each of us. He has performed this task carefully and well, and he did it on time and on budget, setting an example for all future IOM activities.

Kenneth I. Shine, M.D.
President
Institute of Medicine

Foreword

It is a pleasure for me to have this opportunity to introduce this detailed history of the first 25 years of the Institute of Medicine. My involvement with this institution is of relatively recent origin. I assumed office as president of the National Academy of Sciences on July 1, 1993, some 22 years after the Institute was founded. Since that time, I have had the pleasure of collaborating closely with the Institute's current president, Dr. Kenneth I. Shine, as well as with many outstanding members and staff.

As described in this fine volume, the statutory relationships between the National Academy of Sciences, the National Academy of Engineering, and the Institute of Medicine are complex ones, overladen with history. By law, our Academy occupies a special position as the original chartering organization. However, from the beginning of my term, I have tried to ignore these complexities as much as possible. Instead, I like to speak of the Institute of Medicine, the National Academy of Engineering, and the National Academy of Sciences as sister organizations, which work together to provide the expertise that our nation badly needs in the areas of health, engineering, and science. These three "academies" oversee our operating arm, the National Research Council.

Our extremely active organizations carry out many important policy studies at the request of the U.S. government. Altogether, we publish nearly one report dealing with matters of public policy every working day. We are private entities, and the studies that produce these reports are carried out with complete independence from their sponsors. Indeed, it is precisely this independence that gives our judgments their credibility and great value. Modern technologies have recently allowed us to use these reports to produce a large electronic archive of valuable wisdom based on science, engineering, and health that will remain freely available to all through the Worldwide Web at: **www.nap.edu.**

As the importance of science and technology to our society increases, so does the importance and influence of the four organizations that make up the "Academy complex." Ours are not merely honorary membership associations: They are instead service organizations with major responsibilities for directing our nation's future. Our forefathers were indeed wise in keeping medicine, engineering, and science so tightly knit together. Increasingly, we find that the problems we are asked to deal with cannot be answered by any single group of disciplines. Instead, the most important problems must often be addressed by the combined efforts of health professionals, engineers, and scientists. Thus, the close collaboration of the Institute of Medicine, the National Academy of Engineering, and the National Academy of Sciences through the National Research Council allows each of us to serve the nation in ways that are much more powerful than could be achieved by any one institution alone. The current leaders of our three institutions—Ken Shine at the Institute of Medicine, William A. Wulf at the National Academy of Engineering, and I—owe a great debt to the many dedicated people who have gone before us. They have created a truly extraordinary set of institutions that are unique in the world.

In closing, it has been a privilege for me to work so closely with the Institute of Medicine for the past 5 years. The next 25 years will certainly be even more productive, and I look forward with excitement and great expectations to the Institute's many future contributions to the vitality of our nation and the world.

Bruce Alberts, Ph.D.
President
National Academy of Sciences

Author's Acknowledgments

A number of people helped make this project a reality: Mark Santangelo, as chief researcher, did admirable work and provided considerable assistance; at George Washington University, Cyndy Donnell typed the transcripts of the many interviews I conducted and Michael Weeks performed a variety of useful tasks; Harry Marks, of Johns Hopkins University, made many helpful sugestions during the course of the project and passed along many valuable documents; Janice Goldblum and Daniel C. Barbiero, of the Archives of the National Academy of Sciences and National Research Council, went out of their way to make Institute of Medicine records accessible. The IOM History Oversight Committee, particularly its chair, Adam Yarmolinsky, gave both me good advice and helpful encouragement.

Edward D. Berkowitz, Ph.D.

Contents

1

Creating the Institute of Medicine

In the summer of 1964, Dr. Irvine Page, who edited a journal that was widely distributed to members of the medical profession, wrote an editorial on the need for a National Academy of Medicine. With the appearance of this editorial, Irvine Page began a concerted campaign to create the National Academy of Medicine (NAM). Although he was ultimately unsuccessful in creating an entity with this name, he set in motion the forces that would lead to the creation of the Institute of Medicine (IOM) in 1970.

Between 1964 and 1970, key government officials and leaders of academic medicine agreed on the need for an organization concerned with health policy but disagreed over what form that organization should take. Irvine Page had the idea of an advocacy group that would represent the best collective wisdom of the medical profession. In his mind, such a group should be composed primarily of physicians. James Shannon, director of the National Institutes of Health (NIH), thought that the organization's primary mission should be to support scientific research in medicine. Walsh McDermott, a prominent practitioner of academic medicine with a deep interest in public health, believed that the organization should serve as a forum where physicians and other professionals concerned with health policy could work toward the solution of health-related social problems.

Although profound differences divided these three men, they concurred in the belief that the organization should do more than bestow an honor on its members. Each favored an entity composed of working members. Each also realized that the primary reason for forming a new organization lay in the changed relationship between the federal government and medicine. They understood that the federal government, the nation's primary source of funds for the conduct of medical research and the payment of medical care, could not be ignored as a factor in health policy.

The discussions in which Page, Shannon, and McDermott engaged took place in two main settings. Beginning in 1967, Irvine Page organized a series of meetings in Cleveland to discuss the formation of a National Academy of Medicine. These gave way in November 1967 to the activities of the Board on Medicine, a group convened in Washington by the National Academy of Sciences. What emerged in 1970 was the Institute of Medicine. An institute not an academy, the new organization nonetheless reflected the ideas of all three men.

The Page Discussion Group

Between 1964 and 1967, Irvine Page had the field pretty much to himself, although he was not alone in proposing a National Academy of Medicine. In the fall of 1960, a six-person task force, which contained no fewer than four future members of the Institute of Medicine, advised President-elect Kennedy of the need to establish a National Academy of Medicine "comparable to the National Academy of Sciences." President Kennedy showed little interest in this recommendation, preferring to concentrate instead on the creation of Medicare, as the expansion of Social Security to pay the hospital bills of Social Security recipients became known. When Irvine Page wrote his editorial on the need for a National Academy of Medicine in 1964, Medicare was a hotly contested issue. On September 2, 1964, within weeks of the appearance of Page's editorial, the Senate approved a version of the measure.

It was no wonder, then, that Page led his editorial by calling attention to the "important trend linking medicine to government." A group from the medical profession was needed to provide the government with the impartial yet expert advice necessary to make "decisions of wisdom" on questions of medical policy. The American Medical Association, according to Page, approached the federal government with a "grumbling hostility" that limited its effectiveness. The Association of American Medical Colleges, a similarly venerable organization that was in the process of transforming itself from a "congenial 'deans' club' into a powerful lobby for academic medicine," represented the interests of academic health centers, not the medical profession. Few other organizations were large or influential enough to speak for the profession. Page believed that the solution was a National Academy of Medicine, located in Washington, D.C., that would be "truly representative of excellence in all branches of medicine."[1]

Irvine Page was the logical leader of a campaign to create a National Academy of Medicine. Like nearly everyone who helped found the Institute of Medicine, he came from the fields of academic medicine and scientific research. He combined a high academic pedigree (degrees from Cornell and the Cornell Medical School) and scientific prowess with a talent for organization that enabled him to publicize his causes effectively. Born in 1901, he was older than most of the Institute of Medicine's other founding mothers and fathers, more than 65 years of age by the time serious discussions began. His career reflected many of the major developments in American academic medicine. He began by studying the chemistry of the brain and in 1928 received an invitation from the prestigious Kaiser Wilhelm Institute in Munich to begin a department in brain chemistry. After the rise of Hitler made working in Germany uncomfortable, he secured a position at the Rockefeller Institute for Medical Research in New York. These aspects of Page's career illustrated the influence of German models on American science and demonstrated that in the prewar era, support for medical research came as much from private philanthropy as from government grants.

Working at the Rockefeller Institute between 1931 and 1937, Page became interested in the phenomenon of high blood pressure. He both demonstrated the harmful effects of hypertension and showed that the disease could be treated effectively. In 1937, moving to the Lilly Laboratory for Clinical Research, he continued his efforts to "identify and isolate the compounds that affected blood pressure." After the Second World War, he started a research division at the Cleveland Clinic that attracted many topflight doctors and scientists. Not content to limit his work to the laboratory, Page also made rounds at the clinic and became a public health advocate, stressing the importance of exercise and diet in the prevention of heart disease. To further his goals, he helped create such organizations as the American Foundation for High Blood Pressure and the American Society for the Study of Arteriosclerosis.[2]

Irvine Page, then, was far from a typical physician engaged in the full-time practice of clinical medicine; rather, he was a representative of a branch of the profession that enjoyed close relations with academia. Not surprisingly, Page communicated the results of his research by publishing, writing textbooks on the chemistry of the brain, editing books on treatment techniques for stroke and high blood pressure, and producing scientific papers and laboratory reports. Because of his academic prominence and ability to communicate his ideas, he received opportunities to serve in advisory capacities for influential organizations, such as NIH. If he obtained support from the Rockefellers at the beginning of his career, he, in common with many of his colleagues, switched to the National Institutes of Health, whose budget grew from $3 million to $400 million between 1941 and 1960, in the postwar era. Over the course of a long career, therefore, Page had observed the changing relationship between government and medicine at close hand.[3]

At the end of the summer of 1964, Page wrote a letter to his many contacts in the medical profession and in the world of science and asked for their reactions to his proposal to create a National Academy of Medicine. In the letter, he noted that the National Academy of Sciences (NAS) did a good job in representing the interests of science and that engineers appeared well on their way to establishing a National Academy of Engineering within the larger NAS. He wanted to know if a National Academy of Medicine would serve a similarly useful purpose.

Page's correspondents presented him with a bewildering variety of views. Some wanted to wait to see how the National Academy of Engineering turned out. Others expressed skepticism about finding an exact purpose for the organization. Some advocated a close alliance with the National Academy of Sciences, but others felt just as strongly that the new organization should have nothing to do with the NAS. They criticized the NAS as a "self-perpetuating" body. It was a great honor to join, and the "main function . . . is to decide who deserves the honor." "Let us proceed slowly and thoughtfully, but let us proceed," Page concluded.[4] In private, he told a colleague that no one needed to be pushed "but if the thinking about it is open and moves along, the whole thing will gel."[5]

In public, he wrote another editorial for *Modern Medicine* that appeared in March 1965. In this piece, Page again cited the gap between the American Medical Association, which at the moment was leading a last-gasp effort to prevent the passage of Medicare, and the various federal agencies concerned with medicine. Something had to be done to close this gap, because "no serious-minded person denies the role that medicine must play if we plan to have a 'Great Society'. " Summarizing the letters that he had received, he wrote that they demonstrated a need for a National Academy of Medicine and indicated that the material was "certainly there to form it."

In July 1965, Page refined his ideas about the proper membership for a National Academy of Medicine and once again solicited the views of his colleagues. He had no doubt that unlike the democratic and relatively ineffectual American Medical Association, the new organization should draw "from the upper, relatively thin layer of the best medical and scientific and lay talent." There should, however, be no limit on the number of members because the subject of medicine was "constantly growing." Further, membership should depend on a person's capabilities, without regard to degrees or titles. These capable members would be expected to work, not just to participate in "an honorary society for the greater glory of the individual."

For the most part, his correspondents agreed. Dr. Julius H. Comroe, Jr., director of the Cardiovascular Research Institute of the San Francisco Medical Center and later an influential figure in the creation of the Institute of Medicine, argued for the importance of "real working committees picked from individuals best able to give advice on specific problems." Fred Robbins, a Nobel laureate in medicine who worked up the road from Page at Case Western Reserve University and would later be president of the Institute of Medicine, emphasized the need to choose people who commanded respect at the national level and suggested that practitioners of the biological, behavioral, and social sciences be represented. Each of these ideas would influence the subsequent development of the Institute of Medicine.[6]

External events motivated Page to go from ideas to action. Signing Medicare into law on July 30, 1965, President Lyndon Johnson set in motion forces that would lead to a rise in medical expenditures—from 6.2 to 7.6 percent of gross national product between 1965 and 1970— and a fall in the percentage of these expenditures paid by the private sector—from 75 to 63.5 percent in the same five years. Although these trends created opportunities for collaboration between the medical profession and the federal government, they also produced tensions. Similarly, federal support of medical research, although lavish, bred its share of problems among members of Congress, the executive branch, and representatives of medical schools. In June 1966, President Johnson convened a meeting of his top health policy officials and asked them whether "too much energy was being spent on basic research and not enough on translating laboratory findings into tangible benefits for the American people." The mere fact that the President posed the question, according to Stephen Strickland, "fell like a bombshell" on NIH officials and created a ripple of panic among the scientists in medical schools.[7] Not long after this meeting, *Science* printed an item in which it attributed a desire to create a National Academy of Medicine to "reform-minded top officials of the Department of Health, Education, and Welfare." The new organization "would supply the profession with another set of spokesmen and provide the government with a more congenial source of authoritative advice."[8]

Only a short time after President Johnson's meeting with his science advisers, Page used his local connections to secure a grant of $6,000 from the Cleveland Foundation that enabled a group of leading physicians to travel to Cleveland, Ohio and discuss strategy. The meeting took place on January 17, 1967. In addition to Page, 15 physicians, nearly all of whom were on the faculties of medical schools or attached in some capacity to the National Institutes of Health, attended the meeting.[9]

It was Page's show. He set the agenda, hosted the meeting, chaired the sessions, and gave an opening talk, to which he devoted substantial preparation, on the need for a National Academy of Medicine. The new organization, Page said, should be the "voice of moderation, wisdom, and integrity," and it should be "free and beholden to no one" as it "provided advice . . . to any who want to listen." Above all, Page stressed the fact that advice on medical questions could no longer be left to amateurs: medicine's growing scientific base and its increasingly complicated relationship with the government necessitated the creation of a group that could mobilize the best professional opinion in the country.[10]

Although Page dominated the meeting, James Shannon also made his presence felt. Only three years younger than Page, James Augustine Shannon was something of a Washington legend because of his extraordinary success at running the National Institutes of Health. Like Page, Shannon combined interests in science and medicine. He received a medical degree from New York University (NYU) and a Ph.D. in physiology. During the war, he became the director of an NYU research service at Goldwater Memorial Hospital and did work on malaria that led to a Presidential Medal of Merit. After flirting with private industry after the war, Shannon arrived at the National Institutes of Health in 1949 as associate director for intramural research of the National Heart Institute. On August 1, 1955, he became the director of the National Institutes of Health. In this position, he held what Donald S. Fredrickson, himself an NIH director and an IOM president, described as an "uncomplicated philosophy of science" that consisted of "unfettered support of good science and rejection of the bad." Good science tended to mean basic research, rather than research targeted on finding a cure for a particular disease. He blended this uncomplicated philosophy with a sophisticated understanding of congressional relations and of the policy process.[11] Shannon, whose career bore so many similarities to Page's, was just the sort of person whom Page hoped to interest in a National Academy of Medicine.

Addressing the group in Cleveland, Shannon said that a National Academy of Medicine should be able to speak to government and to conduct sound studies. It should be highly professional and not be formed unless it could isolate four or five broad areas of inquiry, because its initial studies would determine its reputation. An overarching purpose of the organization, according to Shannon, would be to "define, enunciate, and promote the health sciences." As a practical matter, Shannon advised that representatives of the group should sit down with Fred Seitz, the physicist who served as president of the National Academy of Sciences, and with Harvey Brooks, a Harvard academic who chaired the Academy's influential Committee on Science and Public Policy, and gauge the NAS's level of interest in a National Academy of Medicine.[12]

A second meeting, held on March 7, 1967, satisfied Page that the need existed for a National Academy of Medicine. Still, he wondered, as had James Shannon, whether the group should proceed on its own or whether it should "come under the umbrella of a chartered organization such as the NAS." In fact, the Cleveland group was already in contact with the NAS. After the first meeting, Shannon and Ivan Bennett had arranged a Washington conference with Frederick Seitz, president of the NAS. They had no trouble getting an appointment. An NAS member, Shannon was one of the federal government's most influential bureaucrats concerned with science. Although a generation younger than Shannon or Page, Ivan Bennett was an academic physician with impressive Washington connections, who had worked at Yale and as the Baxley Professor of Pathology at Johns Hopkins before accepting an appointment as deputy director of the Office of Science and Technology in the Johnson White House.

Talking with Shannon and Bennett, Seitz had offered the possibility of establishing a "board" within the Academy that could begin to do the sorts of studies the Cleveland group wanted and, according to Shannon, "give us a base from which we could begin to operate and to study what sort of organization should emerge." Shannon believed that Page's group should take advantage of the NAS's generosity, even though he thought that there ultimately should be a National Academy of Medicine that was independent of the National Academy of Sciences. Bennett pointed to the advantages of working with a group as prestigious as the Academy. If things turned out badly, the doctors could always back out gracefully and form their own group.

Colin MacLeod, who played a key role in discussions at the second meeting of Page's group, agreed with this assessment, and like the others in attendance, he spoke with considerable authority on the subject of medical policy. He came from an academic and scientific background that was far removed from the daily practice of clinical medicine. A doctor with a strong interest in science, MacLeod had worked at the Rockefeller Institute during the 1930s, just as Page had. From there he had gone on to chair the Microbiology

Departments at New York University and the University of Pennsylvania. In 1963, he took a job in the White House at the Office of Science and Technology, which he later yielded to Ivan Bennett. In 1967, he served as the vice president of medical affairs at the Commonwealth Fund, an important dispenser of funds for medical research and public health. Highly regarded as a scientist, MacLeod, like James Shannon, was a member of the National Academy of Sciences.[13]

Walsh McDermott and the Board on Medicine

As Ivan Bennett noted, the prestige of the National Academy of Sciences, with its federal charter and its proud history that stretched back to the presidency of Abraham Lincoln, was beyond dispute. In 1967, it consisted of about 800 members, each of whom had undergone rigorous nomination and election procedures. Almost all of the members owed their selection to the quality of their published research in the "hard sciences." Upon election to the Academy, a member voluntarily joined one of 18 discipline-specific sections. Physics and chemistry formed the two largest sections. Attempts to increase the number of physicians in the Academy had met with repeated failure. In 1941, for example, Ross Harrison of the Yale Medical School called attention to the "urgent need for the Academy to have in its membership a larger proportion of the distinguished clinicians of the country than at present." He was told that creating a section of medical scientists would result in attracting people who were "merely clinicians," and the idea was dropped for the next 27 years.[14] In the interim, the few medical doctors elected to the Academy were found in the microbiology, physiology, and biochemistry sections. It was not a young crowd; the median age of the microbiologists was 66.5. Very few social scientists belonged to the Academy, and those who did were concentrated in the anthropology and applied mathematics sections. Beyond the prestige that came from joining the Academy, members also enjoyed the benefit of two yearly meetings at which they discussed NAS governance, awarded medals to one another, and listened to scientific papers.

The Academy also contained what observers called an "operational arm" in the form of the National Research Council (NRC). Established in 1916 during the period of mobilization before the country's entry into the First World War, it responded to requests for studies or advice from Congress, the executive branch, and a variety of other private and public sources. The periods of greatest activity for the NRC came during the two world wars. In the Second World War, for example, the NRC's Division of Medical Sciences had advised the surgeons general of the Army and Navy on medical research and other matters related to wartime care. Between 1940 and 1946, the division's advisory committees on war services held more than 700 meetings and 243 conferences and played an important role in shaping the nation's wartime medical policy.[15] In 1946, however, the National Research Council contemplated an end to these emergency activities. The advent of the cold war once again increased the demand for the NRC's services, yet the Division of Medical Sciences initiated little work of its own. As a result of the policy to respond to requests from others, the division contained a bewildering variety of committees, offices, research boards, and panels such as an Office of Tropical Health, an advisory committee to the Federal Radiation Council, the Atomic Bomb Casualty Commission, and the Committee on Radiology. To do its work, the National Research Council drew on the resources of large numbers of scientists in academia and private industry; its reach extended well beyond that of the Academy itself.[16]

Beyond the formal Academy and the National Research Council, the National Academy of Sciences also housed the National Academy of Engineering (NAE). Established in 1963, it marked an outgrowth of the engineering section within the NAS. Engineers felt that the scientists in the Academy failed to accord the engineering profession the respect it deserved. After the engineers received a grant from the Sloan Foundation, the NAS agreed to form what Fred Seitz called "a sister academy under the original NAS Charter." The possibility remained that the National Academy of Engineering would eventually split away from the NAS. By the end of the decade, however, the NAE remained entrenched within the Academy, and the exact relationship between the National Academy of Engineering and other parts of the NAS was a matter of constant negotiation. Engineers continued to feel that they were not first-class citizens of the NAS. Officers of the NAS believed that the engineers elected people to their Academy who were better known as executives of large private firms than as scientists.[17]

Although NAS leaders who served on the Academy's governing council regarded the National Academy of Engineering as less than a hopeful precedent, they nonetheless listened to Colin MacLeod, who reported to them in April 1967 on the events in Cleveland, with considerable sympathy. Seitz told the Executive Committee of the NAS Council that he had already talked with Bennett and Shannon and suggested the creation of an advisory board on medicine. Such a board, Seitz felt, could provide advice on policy questions related to medicine and health and consider the question of whether to form a National Academy of Medicine. For the NAS, it was a way of bringing discussion about a National Academy of Medicine in-house. Furthermore, an ad hoc advisory board of the sort Seitz proposed would involve a minimum of bureaucratic disruption and avoid the time-consuming and often contentious matter of a vote by the membership. Council members agreed with Seitz. If nothing else, creation of an advisory board on medicine would prevent the Carnegie Corporation and the Commonwealth Fund from taking independent action on the matter, as they threatened to do. Seitz felt confident enough about the matter to inform NAS members at their 1967 spring meeting that it appeared "fairly clear" that an advisory board on medicine would be created.[18]

Early in June, the NAS Council approved a motion to establish this advisory Board on Medicine and Public Health. The new Board, charged with the responsibility to "formulate recommendations on matters of policy related to medicine and public health," would report directly to the Council. Seitz told the division heads of the National Research Council that the Board could "possibly lead to the formation of a National Academy of Medicine somewhat analogous to the NAE."[19] The Page group hoped this would be the case and made plans to hold a third meeting.

These plans received a serious setback on June 12, 1967, when Irvine Page suffered what his secretary described as a "mild coronary." Page, the expert on heart disease, now found himself looking at the condition "from inside out." After his heart attack, Page, despite his relative vigor, occupied a less prominent place in the movement to create a National Academy of Medicine. His role changed from that of primary advocate to chief critic of the Board on Medicine.[20]

Seitz realized that the chairman of the Board on Medicine and Public Health would have to be a physician who was also a member of the National Academy of Sciences. His choice was Walsh McDermott, a 56-year-old professor of medicine and public health at Cornell Medical School who recently had been elected to the Academy. McDermott belonged to Page's discussion group, although he had missed both meetings and told Page of his ambivalence toward a National Academy of Medicine.[21] The appeal of his selection lay not only in his familiarity with the movement to create a National Academy of Medicine but also in his unquestioned prominence as a doctor and a public health official.

McDermott was an American aristocrat—"white shoe," in the words of one of his colleagues. The son of a New Haven physician, he attended Andover, Princeton, and Columbia Medical School. During his residency at New York Hospital in 1935, he developed tuberculosis and had to go to the tuberculosis sanitarium in Saranac Lake, New York. Over the course of the next few years he eased his way back into work until he accepted an appointment in 1942 as the head of the Division of Infectious Diseases at New York Hospital. In this post, he performed important clinical trials on penicillin, streptomycin, and other so-called wonder drugs. He also engaged in laboratory work on the effects of antimicrobial therapy on animals, despite the fact that he had never had formal training in either microbiology or experimental pathology. In the 1950s, he shifted directions and became interested in bringing medical treatment to underserved populations, earning substantial fame for organizing a successful tuberculosis treatment program for the Navajos living in Arizona and New Mexico. This experience sparked a continuing interest in public health and led him to become involved in other public health projects both in the inner-city neighborhoods of New York and the developing nations of the world. In the course of these activities, McDermott acquired a plethora of contacts and honors. He served, for example, as a member of the NIH's National Advisory Health Council between 1955 and 1959 and of New York City's Board of Health.[22]

McDermott accepted the appointment as head of the NAS Board on Medicine and Public Health at the end of June. Both he and Seitz hoped that the Board would develop into "a widely respected voice of American medicine." He proceeded to negotiate with Seitz over other members of the Board. They agreed that Ivan Bennett, Colin MacLeod, James Shannon, and Irvine Page, the prime movers behind Page's efforts to launch a National Academy of Medicine, should all be members. Seitz suggested other names, making it a point not to limit membership to physicians. For example, he mentioned Rashi Fein, an economist at Brookings, who had served as a staff member on the President's Council of Economic Advisors and, according to Seitz, "has a deep interest in the social problems connected with medicine," as a potential member of the Board.[23]

Despite Page's illness and the preemptive action of the NAS, the Page group held a third and final meeting on June 28. Colin MacLeod chaired the meeting in Page's absence. Some of the nine physicians present wondered if the Board on Medicine and Public Health would really be able to do anything constructive, because it started from the narrow base of doctors within the NAS. The participants conveyed a sense of regret that Page's group no longer controlled the action on a

National Academy of Medicine. MacLeod urged the group to develop a list of people who should be involved in the Board on Medicine. Those present obliged by coming up with no fewer than 41 names.[24]

The business of constructing a Board on Medicine and Public Health began in earnest on September 13, 1967, at a meeting called by Fred Seitz. Those present included McDermott, Colin MacLeod, Ivan Bennett, and Page. Joseph Murtaugh, chief of the Office of Program Planning at NIH, represented the interests of Jim Shannon. Keith Cannan, who worked for the Division of Medical Sciences of the National Research Council, attended as an NAS staff member. Irving London, chairman of the Department of Medicine at Albert Einstein College of Medicine; Eugene Stead, a professor of medicine at Duke; and Robert J. Glaser, dean of Stanford's School of Medicine, completed the group. All of the people in the room, with the exception of Seitz and NAS staffers, were medical doctors, and each had connections with an academic medical center. It was their task to complete the list of Board members. As Page put it, "Each one of the members appointed to the Board will represent different aspects of social and scientific medicine, such as economics, poverty, etc."

As a practical matter, members of the group agreed on themselves and on Dwight Wilbur, a San Francisco physician who was president-elect of the American Medical Association, and then got bogged down. In the social science fields, for example, the group had only the most cursory knowledge of current practitioners. James Tobin, Milton Friedman, and Carl Kaysen were suggested as economists, even though none of the people in the room was familiar with their work. As for a sociologist, someone in the room offered the name of Daniel Bell, not because of his work in the field of medicine but because he was well known. For the most part, the group concentrated on occupational and demographic categories, such as dentists or people who worked in the pharmaceutical industry, and it acknowledged the need for "at least two Negroes."[25]

The group agreed to ask 22 people to serve, and by the end of September, Seitz reported to the Council that the Board on Medicine "was taking shape." McDermott told Seitz that it was a good list but that it was weighted toward the two coasts and contained almost no private practitioners. Given the people doing the selecting, this was a natural bias. In addition, the Academy had trouble recruiting people from outside the medical profession. A string of distinguished economists declined the offer, as did sociologist Daniel Bell who cited "a lack of competence" on questions related to medicine. Pure scientists approached by the group also were reluctant to become involved with the Board. When Kermit Gordon, the economist from the Brookings Institution declined to be named to the Board, he quickly suggested Rashi Fein of the Brookings staff, whom Seitz had mentioned earlier, as a logical substitute. As Fein recalled, Gordon urged him to accept the position, and he did so with alacrity, pleased to join a "disinterested group that would be able to examine . . . the issues facing the American medical system."[26]

The National Academy of Sciences made a public announcement of the new Board, now called simply the Board on Medicine, on November 13, 1967. In its final form, the Board contained 21 men and one woman (Lucile Petry Leone of the Texas Woman's University's School of Nursing). Two of the Board members were black. Although a majority of Board members were physicians, the group also included two economists, one of whom later resigned, a nurse, a lawyer, an engineer, and at least two people identified primarily as social scientists. Only two of the physicians devoted the bulk of their time to private practice, and one was something of a "ringer." The son of Ray Lyman Wilbur, Herbert Hoover's Secretary of Commerce and a Stanford University president, Dwight Wilbur, although a clinician engaged in private medical practice, nonetheless knew many of the leaders of academic medicine. The rest of the Board on Medicine members either served as medical administrators, did research in academic settings, or both. Twelve held formal academic appointments.

Members of the Board on Medicine, June 1969

Walsh McDermott, M.D.*, †	Joseph S. Murtaugh †
Ivan L. Bennett, Jr., M.D. †	Samuel M. Nabrit, Ph.D.
Charles G. Child III, M.D. †	Irvine H. Page, M.D. †
Julius H. Comroe, Jr., M.D. †	Henry W. Riecken, Ph.D.
John T. Dunlop, Ph.D.	Walter A. Rosenblith, Ing.Rad.
Rashi Fein, Ph.D.	Eugene A. Stead, Jr., M.D.
Robert J. Glaser, M.D.	Dwight L. Wilbur, M.D. †
Lucile P. Leone, M.A.	Bryan Williams, M.D.
Irving M. London, M.D.	Adam Yarmolinsky, LL.B.
Colin M. MacLeod, M.D. †	Alonzo S. Yerby, M.D. †

*Chairman
†Deceased

The appointment of the group made a large splash in the media. The *New York Times* ran a page one story that featured quotations from Seitz and McDermott. Seitz cited the need to apply biomedical knowledge to "critical human needs." He specifically mentioned experimentation on human beings, the role of medicine in attacking rural and urban slum problems, improvement of the quality of medical care, and the reform of medical education as matters that the Board might investigate. McDermott spoke of the "good balanced mix of people who could be counted on for dispassionate and expert judgments about a broad range of problems."[27]

Despite the high hopes, the Board on Medicine faced the difficult task of simultaneously undertaking scientific studies and exploring the feasibility of a National Academy of Medicine. McDermott hoped to build the Board's reputation on the scientific studies it would undertake. For Irvine Page and Julius Comroe, however, the National Academy of Medicine was a constant preoccupation that demanded the Board's total attention. From the beginning, Page worried that the Board would somehow, as he put it, prevent "the evolvement of a free-standing NAM." Although he often contemplated dropping out of the Board on Medicine, he never disbanded his own group of physicians and scientists interested in a National Academy of Medicine, writing reports for them on Board on Medicine activities that, by their mention of confidential discussions, often annoyed McDermott.

"Don't think for a moment that our original group is disbanded," Page told one of his medical acquaintances soon after the creation of the Board on Medicine. He urged fellow Board member Jim Shannon to keep "our ultimate objectives in mind" and informed Shannon that Walsh McDermott "is an extremely capable and nice person but I do not think he is in a class with you." Indeed, Page hoped Shannon would become the first president of a National Academy of Medicine. As these sentiments implied, Page had many criticisms of the way in which the National Academy of Sciences handled the Board on Medicine. He objected to the lack of clarity in the Board's objective and to the mixed nature of its membership. "I am not sure if as unhomogeneous a board as we now have can even begin to fulfill the promises a NAM can make," he wrote Colin MacLeod.[28]

Beyond the matter of a medical academy, the Board faced the problem of meeting high expectations with a group of only 22 people and one professional staff member. He was Joseph Murtaugh, a former staff member at NIH, who had agreed to leave the government and come to work for the National Academy of Sciences. To be successful, the group had to make recommendations that commanded the respect of both government officials and practicing doctors. Yet the group itself contained only one full-time clinician and a few people with recent government experience. Furthermore, the Board hoped to look at the social aspects of medical problems, despite the fact that only four of its members had training in social sciences. Many of the problems on which the Board hoped to focus, such as the plight of inner cities, were large and complex; it would be difficult to define the Board's contributions to their solution.

The Board on Medicine held its first formal meeting on November 15, 1967. All but four of the members attended. Fred Seitz greeted the group and expressed his hope that it might address some of the problems caused by the growing complexity of medicine. Page added his wish that the Board be a place to consider complex problems in a setting free from emotion, professional politics, and other forms of distortion and bias. For all that the Board was' supposed to be a professional tribunal whose members were guided only by the facts, the members inevitably brought their own preferences, formed by their unique experiences, to the discussion. Dr. Dwight Wilbur, whose primary institutional experience was with the American Medical Association, wanted the group to consider the problems caused by the importation of foreign physicians. James Shannon noted the profusion of groups already concerned with medical care and policy and compared the result to a "floating crap game." The Board on Medicine should bring institutional stability to this situation and should make it less of a game of chance and more a matter of putting the best minds in contact with the hardest issues. Adam Yarmolinsky, a lawyer by training with a strong interest in problems related to poverty, noted that his prior contacts with medicine were primarily those of a patient, yet he hoped the Board would address the distribution of medical care in the inner cities. Samuel Nabritt, executive director of the Southern Fellowships Fund who was very involved with questions related to civil rights, agreed that the Board should consider how "to meet immediately and directly the ghetto's needs." The meeting ended with agreement only on the need for more discussion of the types of problems the Board should address.[29]

When the Board reconvened in December, it adopted a seminar format. Board members lectured to one another on such topics as the personal physician delivery system, the functions of the modern hospital, medical education, and the funding of medical research.[30] McDermott believed that these sorts of sessions served the useful purpose of establishing a common language and arriving at a consensus about current conditions that would lead to policy recommendations.[31] Reporting to the members of his group, Page described the tone of the meetings as "pedagogic." "At the moment we are all trying to get our breath, get used to one another, and in *due course* come to grips with these major organizational problems." Page worried, however, that the nonmedical members of the panel did not share his concerns about forming a National Academy of Medicine.[32] It was clear that the Board on Medicine had not yet found its rhythm.

The Heart Transplant Statement

Then, in the second month of 1968, the Board caught a break.[33] On February 2, McDermott brought up the matter of organ transplantation. All of the Board members were aware that only a few months before, the first heart transplant operation had been performed by Dr. Christian Barnard in South Africa. The operation, quickly followed by others performed in the United States, captured public interest at a time when much of the other news, such as the beginnings of the Tet Offensive in South Vietnam, was depressing. Eugene Stead, the Board on Medicine member from Duke, called McDermott and told him that the American Heart Association was prepared to issue a guideline on the use of heart transplants. Stead said that the Board on Medicine would be a much more appropriate organization to issue such a guideline. McDermott reminded Board members that they had already been attacked in the popular press for being inactive. If the Board came up with a statement on heart transplants, it could counter its image. No one appreciated the need for slow deliberation on weighty topics more than McDermott, but as he put it, "we are trying to establish an institution" and certain issues should be seized.

The group held a brief discussion on the subject. McDermott emphasized the fact that heart transplants were experimental procedures that required a careful recording of the operation and its results. Informed consent of the patient should be obtained, and the procedure should be done only on people who would otherwise die. It followed that not all hospitals should undertake heart transplants. When Christian Barnard performed his heart transplant, he lacked, according to Julius Comroe, a proper team of immunologists to assist him. Robert J. Glaser, the dean of the Stanford Medical School that had performed a heart transplant only weeks before, claimed that Barnard had no background in, and little appreciation of, the immunological aspects of heart transplantation, in contrast to Stanford's Dr. Rose Payne, who was a topnotch immunologist. As a result, the radiation in Barnard's operation, according to Glaser, was "given ineptly and in excess." Such practices, the Board agreed, should not be allowed to occur. The Board should therefore caution the medical profession that heart transplants should be done only as an experiment in a proper setting. The public should know that heart transplants were not a therapy that could be relied on until more data had been gathered. Such a statement, Glaser hoped, might serve as a deterrent to adventurous surgeons who wanted to be on the cutting edge of medical practice yet knew little about how to prevent rejection of the transplanted heart. McDermott then turned to Glaser and asked him to draft a statement, with an emphasis on the experimental nature of heart transplants.[34]

Working with Joseph Murtaugh and Walsh McDermott, Glaser prepared a short statement that was ready for circulation by February 20. Instead of waiting for the next Board meeting to discuss the statement, McDermott decided to solicit members' opinions by phone. As he saw it, the statement hinged on two basic principles. Heart transplants could not yet be justified as therapy because no one knew their relative advantages over other forms of therapy. In other words, controlled experiments had not yet been performed, and the public should understand that in the absence of these experiments, heart transplants could not be regarded as a miracle cure. As a second basic principle, McDermott believed that heart transplants should be conducted only in institutions "in which the total array of scientific expertise necessary *for the proper conduct of the whole experiment* can be brought to bear in every case." In other words, heart transplants were more than a matter of surgery; they also involved complicated issues related to immunology. If McDermott's two principles were implemented, the total number of transplants would be small.[35]

Moving with incredible speed, the Board issued what was billed as its first public statement on February 27. The statement polished the rhetoric of the previous discussions. "Cardiac transplantation raises new, complex issues that must be faced promptly," the Board asserted. It could not yet "be regarded as an accepted form of therapy, even as a heroic one." Instead, "it must be clearly viewed for what it is, a scientific exploration of the unknown, only the very first step of which is the actual surgical feat of transplanting the organ." The procedure, in other words, required much more than a surgeon with good hands. Instead, it demanded a "surgical team" that "should have extensive laboratory experience in cardiac transplantation." Furthermore, the "overall plan of study should be carefully recorded in advance and arrangements made to continue the systematic observations throughout the whole lifetime of the recipient."[36]

The media gave the statement prominent play. Walsh McDermott told the *New York Times* that the Board felt comfortable with the statement because heart transplants were "a more clear cut example of all the problems involved in human experimentation" than medicine usually afforded. McDermott said that the Board realized that there were many surgeons who had the necessary skill to perform the operation and who faced a "terrible temptation . . . to embark on the avant garde with . . . the tremendous satisfaction of exercising one's skill." With its statement, the Board hoped to discourage such surgeons. McDermott said that the statement illustrated the Board's mission to consider situations in which medical knowledge outstripped its application to critical human needs.[37]

If the Board wanted to publicize its work, heart transplantation served as the perfect vehicle. The Board used clear and concise language that any layperson could understand to address a pressing matter in a timely fashion. As a consequence, word of the Board's actions spread not just to centers of elite opinion, such as New York and Washington, but also to smaller cities across the nation. Nearly every major newspaper in the country carried a story about the Board's stance on heart transplants. "Don't Try Heart Transplants too Soon, Scientists Warn," read a typical headline.[38]

The statement made a deep impression on intellectuals who followed the transplantation debate and on government officials responsible for funding medical research. In April 1968, the *Saturday Review* printed the statement in full.[39] Senator Lister Hill(D-Ala.), who both authorized and appropriated money for medical research programs, inserted the Board's statement into the *Congressional Record*. In the White House, Joseph Califano, the staff member most responsible for President Johnson's domestic policy, read and commented on the statement. In the Department of Health, Education, and Welfare (HEW), Dr. Philip Lee, Assistant Secretary for Health, called the statement "timely and very appropriate." Wilbur Cohen, about to become HEW Secretary, praised the Board for "bringing a sense of reality and caution into a very complex set of matters." The statement even received international recognition. The State Department's Office of International Scientific Affairs sent the statement to major embassies across the world in the "Friday bag." The United States Information Agency interviewed McDermott for the Voice of America.[40]

Not everyone, even at the elite medical schools, approved of the statement. Some saw it casting doubt on a promising area of medical inquiry. Francis D. Moore, a prominent surgeon at Harvard Medical School, said he was shocked to see such a statement coming from a

Board that contained not a single surgeon or any immunologists who had experience with heart transplantation. He also could not understand why the Board chose to single out heart transplants from the transplantation of other organs. At Brigham and Women's Hospital, a Harvard teaching hospital, surgeons had done their first kidney transplant in 1951, although the procedure had involved identical twins and posed no immunological problems. Ignoring the fact that Glaser had contacted the members of his faculty who had performed heart transplant surgery, Moore thought it would have been prudent "for the National Academy of Sciences to seek some consultation from those who have been intimately concerned with these problems for almost twenty years." He felt that the Board did not make it sufficiently clear that heart transplants could "rest secure in the scientific, ethical and moral climate which has already been characteristic of the transplantation of other organs in the United States."[41]

In a sense, Moore accused McDermott of exploiting the interest in heart transplants in an opportunistic manner. In reply, McDermott argued that the statement was intended to be a positive one, not meant to eliminate heart transplants but rather to make sure that such procedures were performed in places that had the "capability to make all the relevant observations." Most of the cardiac surgeons with whom McDermott had discussed the statement regarded it as "sensible and helpful," setting down guidelines that they themselves followed.[42] The statement, according to McDermott, exemplified the Board's mission of acting as a "disinterested group" that sought "to be helpful on the more important issues as they arise considering both medicine itself and its relationship to our society."[43]

Still, the heart transplant statement, for all of McDermott's lofty ideals, possessed elements of self-interest that were difficult for members of the Board on Medicine to perceive. As a pragmatic matter, the statement counted as a great success. As Irvine Page noted, the Board was "well ahead of the game" and could only hope to do "as well next time."[44] It did not trouble either Page or McDermott that the statement favored the types of institutions with which they were associated over other institutions; they simply assumed that their institutions were superior. Robert J. Glaser, chief author of the heart transplant statement, worked for a medical institution that had already performed such transplants, one whose scientific expertise, according to the Board on Medicine, should allow it to continue to do these operations. To put it more bluntly, the head of the Stanford Medical School had helped draft a statement saying that complicated operations such as heart transplants should be done only in places like the Stanford Medical School. No one on the Board on Medicine was likely to challenge such a statement.

Business of the Board

The heart transplant statement gave a sense of momentum to the operations of the Board on Medicine. As if in direct response, the Commonwealth Fund announced that it would give $150,000 to the National Academy of Sciences to support the Board on Medicine. It marked the start of what would be a long-standing philanthropic relationship between the fund and the Institute of Medicine. In awarding the money, the head of the Commonwealth Fund spoke of the "urgent need for an institutional entity of authoritative standing to serve as an independent, objective resource for dealing comprehensively with the problems and decisions confronting medicine and health in American society."[45] Even when the hyperbolic foundation rhetoric is discounted, such a statement showed the hope that the Board engendered in elite medical policy circles.

Although a lack of funds would become one of the Board on Medicine's and the Institute of Medicine's chronic problems, it looked at first as if raising money would be easy. The Board, unlike other voluntary organizations that served a social purpose, enjoyed the backing and prestige of the National Academy of Sciences. Its mission, although vague, appeared to be related to America's social problems at a time when the mid-1960s optimism about solving these problems through Great Society programs had begun to wane. The combination of pedigree and relevance was very attractive to private foundations. As if this were not enough, the Board had some direct connections to the foundation world. Colin MacLeod, the same person who pleaded the Board's case to the NAS Council, worked for the Commonwealth Fund, which became an early supporter of the Board. Many Board on Medicine members served on foundation boards, and some, such as Walsh McDermott and Robert J. Glaser, went on to play major roles in the foundations that supported medical research. In Jim Shannon, the head of NIH, the Board maintained a connection to the largest supporter of medical research in the world. Although Shannon left NIH in 1968, he nonetheless kept the Board abreast of NIH and other government activities.[46]

Colin MacLeod took a direct role in the Board's outreach to the philanthropic community. He arranged a meeting at the Commonwealth Fund in New York that allowed Seitz, McDermott, and Murtaugh to discuss the Board's work with representatives of five large foundations. McDermott looked to the foundations represented at the meeting, such as the Carnegie Foundation and the Rockefeller Foundation, to provide core support for the Board. He wanted to increase the Board's staff by at least three professional staff members. At the end of the meeting, Margaret Mahoney of the Carnegie Corporation took McDermott aside and told him that her foundation would be very interested in participating in a study of urban health services. She also told McDermott that the foundations worried about the widespread dispersal of their funds, which limited their effectiveness. She advised McDermott to get several foundations to work in concert to support the Board on Medicine.[47] The Board on Medicine followed her advice and secured initial backing from Commonwealth, the Rockefeller Foundation, the Milbank Memorial Fund, Carnegie, and the Association for Aid to Crippled Children.[48]

A good part of the attraction for these foundations was the fact that the Board worked on social, and not merely medical, problems. If the Board needed a reminder of the urgency of America's social problems, it received it during its April meeting. Originally scheduled for the downtown offices of the National Academy of Sciences, the meeting took place in the suburbs to avoid the disruptions caused by the riots that followed the death of Martin Luther King, Jr. At this meeting, the Board faced the fact that despite the heart transplantation statement, it had still not initiated any substantive studies. In partial response, members decided to create two study panels. One, on medical education, came at the urging of Jim Shannon, who sought nothing less than a major redesign of medical education to incorporate changes in science into the curriculum. McDermott asked Ivan Bennett to head a five-person subcommittee charged with developing a proposal. The other study panel, on health services, had the daunting task of examining "existing programs, mechanisms, and laws" that related to the delivery of health services and of somehow formulating "a creative diagnosis" of the issues in the field. McDermott chose economist Rashi Fein to head this five-person panel.[49]

Both panels made initial progress. In June, the Board gained NAS approval to enter into a $40,000 contract with the National Institutes of Health for the initial phase of a study of medical education.[50] The Board also refined its proposal in the area of health services, concentrating on the topic of health and the poor. It proposed to define the problem by asking if the health conditions of the poor were different, to examine how the poor were being served, to analyze government health programs that served the poor, to project the future medical needs of center-city dwellers, to test the effect of income transfers or vouchers on improving access to health care, and to design government programs that might better serve the poor. If the project lacked focus and appeared to be overly ambitious, it nonetheless represented a tangible product of the Board's long discussions.[51] By the fall, a Board Panel on Health and the Disadvantaged, chaired by Samuel Nabrit and containing five other Board members and four outside experts, had met in Washington for further consideration of the study.[52]

Despite this semblance of progress, the Board faced many internal and external challenges. In the first place, it drew criticism from within the National Academy of Sciences. One epidemiologist, a member of the NAS since 1948, questioned whether the Academy was qualified to appoint a Board on Medicine. He noted that the Academy itself still did not have a functioning Section on Medicine (although one would be formed in 1968) and that Board members were not "primarily people who have been involved in the provision of medical care to the public." The Board's social mission, so attractive to outside funders, also drew criticism from inside the Academy.[53] Second, the very attractiveness of the Board's mission, combined with the slow pace at which it moved, invited competition from other entities that claimed to speak in the public interest. Senator Walter Mondale (D-Minn.), for example, hoped to create a Commission on Health, Science, and Society to "study and evaluate research in medicine." Such a commission could provide competition for the Board.[54] Third, the Board still had to deal with the question of forming a National Academy of Medicine.

Neither Julius Comroe nor Irvine Page would let go of this issue. Comroe argued that the Board was placing too much emphasis on its social mission and too little on forming a National Academy of Medicine. Suggesting that the Board change its name to the "NAS Board on Social Medicine," he thought another group should organize a National Academy of Medicine. Comroe was worried that the Board spent its time on large and intractable problems, such as the conditions in urban and rural slums, that required the entire nation's attention and to which the Board could contribute little. Comroe, a blunt man, did not hide his feelings from Walsh McDermott. "I think that it is very nice for 22 men in a wide variety of disciplines to meet once a month and move at a relatively slow pace toward solution of major problems," he told McDermott, "but the work could go very much faster . . . if a large prestigious group, more representative of medicine were formed to do the work. . . ." Page agreed with Comroe. In Page's opinion, the Board had become "the McDermott Committee of the NRC" and he hoped that Comroe, Colin MacLeod, and Jim

Shannon could save the organization from "becoming a small, self-satisfied outfit which can never properly undertake the problems I think we all had in mind when we got the thing started."[55]

Debate over a National Academy of Medicine

By July 1968, McDermott believed that it was time to begin a serious discussion on whether to form a National Academy of Medicine. He arranged a special one-day meeting to consider the matter and asked Comroe and Page to lead the discussion. In a paper prepared for the meeting, Page critiqued the Board on Medicine as too small, too narrow, and too concerned with the socioeconomic dimension of medical problems. With the socially minded Board on Medicine and the scientifically minded National Research Council, medicine found itself in a "schizophrenic state," a "split appendage" of an organization in which it had "no voice." Page concluded that a National Academy of Medicine should be formed, with or without the help of the National Academy of Sciences.[56] Comroe, for his part, prepared for the discussion by asking the opinions of 20 prominent physicians. All 12 of those who responded believed that there should be a National Academy of Medicine, although they differed on whether it should be affiliated with the NAS. Comroe emphasized that his correspondents felt a National Academy of Medicine should be organized by people actively engaged in medical practice, education, research, or administration. As for his vision of a National Academy of Medicine, Comroe thought it should have a permanent home, a full-time president, and at least 20 members in full-time residence, many of whom would be on sabbatical.[57] Not surprisingly, he saw the NAM as a place to which academics would come to write and do research.

From the moment of Comroe's and Page's presentations on a National Academy of Medicine, the Board on Medicine concerned itself with little else. At the end of six hours of conversation, McDermott said he detected neither a consensus in favor of an academy nor one against it. He decided to appoint a subcommittee to investigate the matter. Headed by Irving London, a summa cum laude graduate of Harvard and a Harvard-trained doctor who chaired the Department of Medicine at Albert Einstein College of Medicine, the group contained six other Board on Medicine members. Because the group included Shannon, Comroe, Wilbur, and Page, it was comprised, as Comroe conceded, of "a very clear majority dedicated to the formation of an academy." McDermott served in an ex officio capacity.[58]

This group, with the unwieldy title of the Ad Hoc Panel on Further Institutional Forms for the Board on Medicine, met at the end of September 1968. The political climate lent a sense of urgency to the proceedings. When the Board started its work, the Democrats were in firm control of the White House and Congress. It now appeared as if Republican Richard Nixon might win the election and federal support for medical research and for social welfare programs, such as Medicaid, might weaken. As Adam Yarmolinsky told the Board, "Whoever is President of the United States—Humphrey, Nixon, Wallace, it is less likely that the kind of advice that we have to offer which involves spending money and doing things will be accepted."[59]

Whatever the urgency of the political situation, the Ad Hoc Panel moved at a very deliberate pace. A first meeting served only to set the stage for a second. "The discussions were quite tentative," Joseph Murtaugh reported to Fred Seitz, "with a great deal of uncertainty being reflected by the members in respect to many important points."[60] One of Page's allies wrote that he sensed a "foot-dragging attitude" on the part of McDermott. He urged Page to act independently of the National Academy of Sciences. He suggested that if nothing happened in the next few months, Page's original committee should be reactivated and a National Academy of Medicine should be organized on an independent basis.[61] The head of the Dartmouth Medical School told Page that he was "bullheadedly holding to my original bias" in favor of a National Academy of Medicine. "The more I have thought about it . . . the more I feel the academy is the only position to take," Page replied. He attributed opposition to the NAM to "public health people" and those "concerned with economics" who believed that "if medicine had the upper hand, we would all revert to our antediluvian days of which they are so highly critical."[62]

Walsh McDermott, whom Page would have classified as one of the public health people, refused to be stampeded. In a letter to London he tried to set the intellectual tone for the discussion. He began by pointing out that everyone agreed on the need for an institution, free of special interest, that included nonphysician professionals as well as physicians. The disagreement came over what purpose the institution should serve. The Page and Comroe faction favored an organization that spoke *for* medicine; McDermott wanted one that spoke *about* medicine. The first concept involved a group that addressed problems within medicine and expressed the medical viewpoint. The second concept addressed the need for an entity that "would speak to the issues from the position of all-round competency rather than one that would speak from within medicine." This concept reflected McDermott's belief that the major problems facing medicine extended beyond the professional boundaries of medical doctors. Because McDermott admitted that he favored the second concept, he came down on the side of expanding the number of people on the Board on Medicine without changing its basic shape or character, rather than creating a National Academy of Medicine. It seemed to him that the Board represented the best hope of interdisciplinary collaboration, in which physicians could work with social scientists on such problems as the reform of medicine and medical education. An Academy of Medicine would develop a tendency to be overprotective of the medical profession and to shy away from the most critical problems.[63]

In a rebuttal to Walsh McDermott, Page charged that McDermott had stacked the intellectual deck. McDermott assumed that physicians would act in a biased and defensive manner without any evidence to support those assertions. Page argued that McDermott's was "the counsel of timidity" that expressed a "lack of faith in the competence of selected physicians." Physicians, according to Page, must have a "strong voice in determining the destiny of medicine," and anything less was "a weak compromise for which interest will soon flag."[64]

The next meeting of London's committee took place in the midst of considerable confusion. First, the presidential election turned out as many Board on Medicine members had feared. Second, the National Academy of Sciences was undergoing its own change of administration. Fred Seitz would leave at the end of the fiscal year (July 1969), and the Board would have to deal with a new NAS president. In addition, the NAS was forming a section on medical sciences that could conceivably sap some of the energy from the campaign to create the National Academy of Medicine. Because these changes made the future seem less than clear, the second meeting of London's committee, like the first, ended with little substantive agreement except that each member would write down his preferences about a new institutional form for the Board on Medicine and send the results to Dr. London.[65]

When the full Board on Medicine convened on November 20, 1968, London could offer his colleagues only a status report on the deliberations of his subcommittee. Despite the uncertainty, some ideas appeared to be taking shape. Membership in whatever organization evolved should be on a short-term basis. Each of the members would be expected to work; no one wanted a purely honorific organization. There should be a full-time staff, headed by a full-time officer or officers. The majority of the people on the Ad Hoc Panel appeared to be heading toward the use of the word "academy" in the organization's title. Most people agreed that the Board on Medicine in its present form had failed to meet the expectations generated by Page's discussion group and other advocates of a National Academy of Medicine. Still, profound differences remained between the Page–Comroe faction and McDermott. As Comroe put it, "Some of us believe very strongly that if it is to be a Board on Medicine or an Academy of Medicine, then it must be predominantly made up of medical people." If such a board or an academy were to be accepted, it had to be accepted among medical practitioners. This meant it had to overcome the stereotype that physicians always acted in their self-interest, desired nothing more than a trade union, and exhibited no concern for the country's future.[66]

Irvine Page aptly summarized the situation at the end of 1968. In an unguarded moment, he wrote that "there have been many, many arguments over the past year and a great deal of soul searching going on and there is no way of knowing how it is going to come out." To Dwight Wilbur, Page's friend and ally on the Board, it appeared that Walsh McDermott was "doing his best to fan the flames in his direction or on his behalf."[67]

As Wilbur's comment suggested, the Page faction spent much of the next year and a half in a state of frustration. One source was the inevitable delay. Another, more enduring source of frustration came with the announcement that Philip Handler, chairman of the Department of Biochemistry at Duke, would become president of the National Academy of Sciences on July 1, 1969. Unlike Seitz, Handler held no particular brief for a National Academy of Medicine. High on his list of priorities was a desire to strengthen the life sciences within the Academy. As part of this agenda, he hoped to elect more practitioners of medical sciences to the Academy and to strengthen the management of the National Research Council. None of this meshed with Page's plans.[68]

McDermott, too, experienced his share of aggravation. As the meetings of the Board on Medicine became more contentious, his job became more difficult. His finely honed sense of obligation did not permit him to retire from the Board until he had settled its future. This commitment forced him to attend endless rounds of meetings with Board members and NAS officials. Because McDermott regarded it as his duty to serve as an honorable intermediary between the Board and the NAS, he was often the bearer of bad news.

Nor did the Board's research projects bring much encouragement to Page, McDermott, or anyone else. The project to reform medical education gradually unraveled. It had been tacitly agreed with Jim Shannon that the National Institutes of Health would support the project. On the strength of the initial $40,000 from NIH, the Board recruited a study director and assembled a number of panels. By the beginning of 1969, it had become clear that NIH had developed different priorities since Shannon's departure. The study director resigned, and efforts began to secure funds from other sources. The project on medical services for the poor experienced a similarly rocky start.[69]

London's panel to decide on a National Academy of Medicine appeared to be the only Board on Medicine activity that mattered. The group met for a third time at the end of February in Palo Alto, California.[70] This time the group managed to get past its previous impasse and reach a decision. The group had already decided on the need for an "institutional framework in medicine" that could confront medicine's problems. The questions centered on what form such an institution should take. The panel debated three options: (1) continuation of the Board on Medicine, (2) transformation of the Board on Medicine into a National Academy of Medicine under the Charter of the NAS, and (3) creation of a free-standing National Academy of Medicine. In the end, the panel decided on the second alternative. It recommended that the "Board on Medicine take the steps necessary to secure the creation, under the Charter of the National Academy of Sciences, of an autonomous National Academy of Medicine." The panel even included a timetable that culminated in the formal creation of the National Academy of Medicine by July 1, 1970.[71] Joseph Murtaugh hastily drew up a proposal for a National Academy of Medicine that included members who would serve for no more than 10 years; it would have an initial membership of 250 people, a balance of three to one between medical professionals and others, and a full-time elected president.[72] Although Murtaugh could not have known it, he had, in fact, written the first draft of a plan for the Institute of Medicine.

At first glance the London panel's report appeared to be a victory for Irvine Page, who favored the creation of a National Academy of Medicine, and a defeat for Walsh McDermott, who opposed it. As always, however, it was the details that mattered. If London's group had its way, the National Academy of Medicine would be part of the National Academy of Sciences, a prospect more congenial to NAS member McDermott than to Page. Nor did the panel specify how much of a social, as opposed to a medical, content the new academy would have. Therefore, although Page felt a cautious sense of optimism, he warned Dwight Wilbur that "our troubles aren't over."[73]

The Board on Medicine debated London's report on March 12, 1969. In his presentation to the Board, London emphasized the advantages in terms of prestige and financial support that would come from linking the NAM to the established National Academy of Sciences. When Philip Handler dropped in on the meeting, he too argued against what he described as a free-standing academy. "I am afraid it would be thought of in the world at large as a guild of some sort," and it would be encumbered with the "stigma that guilds have in the world," he said. McDermott agreed that "the greatest single trap in which this campaign can be caught is guildism."

Listening to the discussion with relative complacency, Board members voted to accept the panel's main recommendation. With a unanimous vote, the Board put itself on record as favoring an autonomous National Academy of Medicine under the National Academy of Sciences. In Comroe's account of the meeting to the absent Irvine Page, "there were only a few feeble arguments opposing the proposal for a National Academy of Medicine and these were dealt with very ably by those around the table." During the rest of the meeting, the Board attended to some of the details. If the Council of the National Academy of Sciences approved the proposal, then the president of the NAS would appoint an organizing committee to write what amounted to a constitution for the new organization. The Board on Medicine would then be expanded to 50 members who would select the first members of the National Academy of Medicine. Within three years of the founding of the National Academy of Medicine, three-quarters of the members would come from medicine and the medical sciences, and one-quarter would come from the social sciences, administration, law, and engineering.[74]

Rejection by the National Academy of Sciences

Once again, the campaign to create a National Academy of Medicine accelerated. Walsh McDermott presented the Board's plan to the Executive Committee of the NAS Council at the end of March. The NAS officers asked him a lot of pointed questions. They wondered just how members would be chosen, how the new academy might develop a working relationship with the National Research Council, and whether the NAM might take positions on public controversies that would put it at odds with the impartial National Academy of Sciences. In the end, the Executive Committee "accepted in principle the concept of a National Academy of Medicine" and agreed that the full NAS Council would take up the matter in June. However, preliminary discussions in the Council, conducted behind closed doors, showed that many NAS Council members remained skeptical of the venture. They questioned the need for a new academy's autonomy. As far as Philip Handler and the NAS Council were concerned, the matter was far from settled.[75]

In the course of the May discussions of the Board, Walsh McDermott confided that he was having his own crisis of confidence. Although he would do his best to argue the case for a National Academy of Medicine to the Council in June, he still had doubts about the wisdom of this approach. Two things were essential to him—that the NAM not be solely honorific and that it contain a mixture of disciplines. "I have serious questions that anything that is called an academy is the proper instrument to meet these two purposes," he said. Irving London, who sympathized with McDermott's position, explained why he nonetheless favored a national academy. If the Board followed McDermott's advice and established something like a "National Medical Council" or the "National Board on Medicine," then "quite another group in this country would come along calling itself the National Academy of Medicine and would create enormous confusion in the public mind."[76]

Setting his doubts aside, McDermott issued the final assignments for the presentation to the NAS Council. He, London, and Adam Yarmolinsky would draft a document that would be presented to the Council. A delegation from the Board on Medicine comprised of McDermott, Glaser, London, and Yarmolinsky would make the formal presentation. There was nothing surprising about these assignments, except perhaps for the inclusion of Yarmolinsky. His selection reflected the Board's growing dependence on Yarmolinsky as a draftsman and, because he was quick thinking, as a spokesman. Yarmolinsky became the Board's informal counselor, advising the group on legal matters and matters of organizational design. His experiences in the Kennedy and Johnson administrations made this Harvard law professor the closest thing the Board had to a Washington insider.[77]

Jim Shannon, the Board's consultant and other certified Washington insider, took the case for a National Academy of Medicine to the public. Speaking in Chicago, Shannon told the American College of Physicians that the National Academy of Medicine would be a superagency that would reorganize America's "nonsystem" of health care. It would do for medicine what the National Academy of Sciences had done for science and technology. Addressing a Travelers Research Corporation seminar in Hartford, Connecticut, Shannon said that the members of the National Academy of Medicine would be broadly informed and "objective enough to give true leadership." *Nature* reported that there was a good chance that a National Academy of Medicine would exist by 1970 and that Shannon "seems to have been the chief instrument in planning the new academy."[78]

Shannon joined the Board's representatives when they made their pitch to the NAS Council on June 7, 1969. The two-hour session resembled a Supreme Court hearing, with NAS Council members taking the part of judges. Scientists on the Council peppered the physician and lawyer who stood before them with questions. Would it not be better to expand the Board on Medicine? How could the National Academy of Medicine ensure that its members would be of high stature? Would the inclusion of medical practitioners debase the status and prestige of the new academy?

Although Shannon and the others did their best to respond, the answers seemed only to create more questions. The Council should not look on an Academy of Medicine as a threat or as competition, argued Shannon, but rather as a "complementary institution necessary for the further evolution of science." It would be a source of disinterested advice. This led George Kistiakowsky, the Harvard chemist and NAS vice-president, to wonder if offering such advice would make the new academy a lobbying organization and threaten the NAS Charter. Phil Handler also stressed the differences between the two sorts of academies. The National Academy of Medicine would initiate views on public policy rather than limiting itself, in the traditional NAS manner, to responding to requests for advice. The broad scope of the NAM mission also troubled the scientists, who wanted the disciplinary borders of the new academy to be clearly drawn. At the end of the discussion, Alvin Weinberg, an applied physicist who worked at the Oak Ridge National Laboratory, said that he would have a difficult time voting for the proposal because of what he called a substantial "impedance mismatch" between "some of the concepts set forth for the NAM and those which have been traditional for the NAS."[79]

When the Board on Medicine representatives left the room, Council members expressed serious misgivings about establishing a National Academy of Medicine. Handler said that demands to create separate academies would proliferate and soon the social scientists would ask that they too have an academy. The National Academy of Sciences might then become so diluted that its prestige would disappear. Wallace Fenn, a physiologist who worked at the University of Rochester's medical school, thought it would be much better to expand the Academy's Medical Sciences Section and the NRC's Division of Medical Sciences than to create a new academy. Council members agreed that the Board's proposal should be examined with "great care and deliberation." Handler, who was about to take over as NAS president, proposed—and the Council agreed—that action on the proposal be put off until the Executive Committee meeting in July.[80]

Soon after the June meeting, Handler wrote the NAS Council a letter that further revealed his doubts about a National Academy of Medicine. He realized that there was a mismatch between the qualifications of the distinguished scientists elected to the NAS and the practical expertise needed to respond to requests for advice on policy questions related to science. The Academy had too few clinical medical scientists, applied agricultural scientists, engineers, and social scientists. This put many pressures on the Academy, such as the request to form the National Academy of Medicine. The advocates of an NAM wanted to "undertake a type of lobbying activity which is not in keeping with the history of the Academy" and to have a "mixed membership which would, in effect, make them independent of this Academy for advisory functions." If the federal government turned to the NAM and similar entities for advice, the NAS would "be left with only minor advisory functions. Its role in our national life will have been very markedly diluted and its prestige very significantly eroded." Handler proposed to create a Division of Medical Sciences, one of five such divisions in a reorganized NAS, that could elect its own members. Such a division might preempt the campaign to create a National Academy of Medicine.[81]

Although Joseph Murtaugh called the Council's reception of the Board on Medicine "cordial," the proposal for a National Academy of Medicine came at a sensitive time in NAS history.[82] The resulting dispute reflected in part the differences in outlook between the hard and applied sciences. The physicists, chemists, and mathematicians who made up the bulk of the National Academy of Sciences regarded most social problems as too imprecise to define in a meaningful way. Because social problems did not lend themselves to the formulation and testing of hypotheses, they could not be solved through the scientific method and hence were not the legitimate business of the National Academy of Sciences. The economists and sociologists who might have disagreed with this proposition were too few in number in the NAS to influence its outlook.

Both McDermott and the Page contingent differed in their views from NAS scientists. For McDermott, the social aspects of medicine were the most important things for a National Academy of Medicine to address. Shannon, Page, and Comroe, each of whom became NAS members, were closer to the outlook of pure scientists. They, too, had doubts about the social aspects of the Board on Medicine's work—in part because they came from the world of medical research, rather than the fields of public health, health services, or health care financing. Still, they believed that an effective National Academy of Medicine required participation from a broad cross-section of the medical profession. This put them at odds with the scientists who regarded an academy as a place for researchers, not mere clinicians; for thinkers, not policy activists. Furthermore, the scientists at the National Academy of Sciences, led by Handler, were protective of the Academy as a place where merit alone secured membership. In the Nixon era, with both academia and the undirected nature of scientific research under attack, preserving the Academy as a sanctuary from ephemeral politics seemed all the more important.

When the Executive Committee of the NAS Council met on July 19 in the Academy's facilities at Woods Hole, Massachusetts, to make a final decision on the Board's proposal, Handler's views prevailed. Although the NAS was generally sympathetic to the objectives of the Board proposal, it objected to the issuance of policy statements on political questions and to the "variable criteria" for membership. The Executive Committee hoped to work toward increasing the number of NAS members and, in doing so, to meet many of the Board's objectives. If the Board wished to proceed with an independent NAM, it should feel free to do so but without the help of the NAS. Finally, the fact that engineers had gained their own academy should "not be followed as a precedent."[83]

On the following day, McDermott and his group met with the NAS Executive Committee and learned of its decision: the NAS Council had rejected the concept of a National Academy of Medicine. Any hope of achieving the Board's objectives would have to occur in a reorganized NAS. In the meantime, the Executive Committee advised the Board to continue with its activities and to have faith in the committee's goodwill. Somewhat stunned, Yarmolinsky and London asked how the movement to create an independent National Academy of Medicine could be contained as the NAS reorganization process proceeded. There could be no way of preventing this from happening, Handler replied. He thought that in the long run, however, the prestige and status of the NAS would prevail.[84]

Handler tried to put a positive spin on the Council's actions. He emphasized that the NAS Council "warmly" approved of the Board's objectives. The Council's differences with the Board came over the mechanism of a National Academy of Medicine, which was "not welcomed by the Council at this time." Rather, the Council proposed to restructure the NAS, and whatever structures emerged would involve "increased membership of clinical investigators and of individuals in other disciplines related to the problems of health care." Handler hoped that the process could be completed by the following spring. He concluded by thanking the Board for its proposal, which "has heightened our consciousness of the need for early reforms of our structure and mechanisms."[85]

The Diplomacy of Reconciliation

At a July postmortem meeting of the Board, Page said that the movement to start an independent National Academy of Medicine could no longer be contained. Dr. London believed that the group was the victim of timing. If Fred Seitz were still the NAS president, things would have turned out differently. McDermott tried to move beyond the anger and get the Board to decide on what to do now. Should the Board continue, or should its energy be put into forming an independent National Academy of Medicine? Dr. Bryan Williams, the physician from Dallas, said he did not want to wait any longer; it was time to form a National Academy of Medicine. Dr. Comroe urged the group to go back to Page's original conception of a National Academy of Medicine. Ivan Bennett agreed that the group had lost both time and money by working with the National Academy of Sciences. For all of the anger, however, the group could still not make a clean break from the NAS. Instead, McDermott was instructed to tell Philip Handler that it looked as though "the road is going to be that of an Academy of Medicine, that it is our hope that this can be done in a way that represents a natural disengagement from NAS with a lot of help from NAS." McDermott, meanwhile, clung to his hope that something could be worked out with the National Academy of Sciences.[86]

The situation had turned out as Page had thought. "It looks now as though the NAM is going to be formed independent of the NAS," he told a friend, "We are wiser but just about where we left off when the meetings were transferred to Washington." It was time to reunite the Cleveland group and get down to business. On August 15, Page sent a form letter to the leaders of his original Cleveland discussion group in which he announced tentative plans for another small meeting. The whole experience, he explained to Jim Shannon, demonstrated that a "chairman unsympathetic with the purposes of a group can really hamstring the progress of a committee."[87]

It was not a good time for the Board on Medicine. Within a week, it became clear that the study of medical education, which had received a final rejection from the National Institutes of Health, would be disbanded. The staff hired for this study began to put their resumes in the mail. With the loss of this project, the rationale for continuing the Board on Medicine looked even weaker.[88]

In this deteriorating situation, Handler and McDermott engaged in a confrontational form of diplomacy. McDermott advised Handler that at least 14 members of the Board doubted whether an accord could be worked out with the Academy. Some of these, however, still believed that it would be more worthwhile to maintain some sort of link with the Academy. Handler informed his executive board that the Board on Medicine had decided to go ahead with the initial planning for a National Academy of Medicine and wondered what, if anything, the NAS should do in response. Members of the Council's Executive Board told him to hold firm. George Kistiakowsky recommended that the NAS do nothing to oppose the creation of a National Academy of Medicine. It should, however, insist that the Board undertake this activity on its own time or through other organizations. Wishing the Board well, the NAS should go about its business of increasing the number of physicians in the Academy and restructuring the National Research Council. Encouraged by this advice, Handler did not appear willing to make many concessions to the Board. "I feel rather strongly that, in the end, creation of an independent National Academy of Medicine is not in the national interest," he told Joseph Murtaugh. The National Academy of Sciences would not be blackmailed by the threat of an independent National Academy of Medicine. At this point, the situation resembled the sorts of power struggles that scientists were beginning to witness over such things as the creation of black studies programs on their campuses.[89]

As the summer continued, both sides retreated from their harsh positions and began to engage in pragmatic bargaining. Walsh McDermott did all he could to soften Handler's position through a series of phone calls and telegrams. Working closely with McDermott, Handler drafted a letter in which he offered McDermott a chance to participate actively in the NAS committee charged with reorganizing the National Research Council. He also raised the possibility that a series of what he called "institutes" might be formed within the National Academy of Sciences, "one of which might well be an institute on medicine or health affairs." As he made these suggestions, Handler, along with the Executive Committee of the Council, urged that "the Board not take precipitate action at this time with respect to the planning of an external Academy of Medicine." Instead, all alternative plans should be examined and an "acceptable accommodation" found. In this manner, health affairs could assume "a suitable, highly visible, and decidedly dignified position within this total body" and the fragmentation of the NAS could be avoided. "Your . . . letter to me was splendid," McDermott replied.[90]

In September 1969, negotiations continued at a 12-hour meeting attended by Handler and all the principal players from the Board on Medicine, a sort of executive committee. Board on Medicine participants tried to determine which of their conditions were nonnegotiable. Merely adding members to the Board, they agreed, was unacceptable. At a minimum, they required a large institution with freedom to develop its own program, a diverse membership, and the word "academy" in the title. Handler, for his part, tried to be as conciliatory as he could, yet he admitted that the issue of a title for the new medical organization still presented considerable difficulties. He once again mentioned the possibility of an institute and hoped that this might provide "a sufficiently prestigious image." He also speculated that the entire NAS might be reorganized to contain five different subacademies, including the Academy of Medicine. Because such an academy (or subacademy) would maintain the NAS's membership standards, only scientists would be allowed to join. This left no room for the mixed membership coveted by McDermott. As for a completely independent National Academy of Medicine within the NAS, Handler saw no possibility of this. "If the Board insists on going down that road, we are at an impasse," he said.

After listening to Handler, the exhausted group identified three options, each of which meant giving up something. One was to proceed with an independent NAM and abandon the discussions with the NAS (the Page solution). Another was to find a place for medicine within a reorganized NAS structure and abandon McDermott's notion of a group with a mixed membership (the Handler solution). A third involved creating a new institution within the NAS, with a mixed membership, and abandoning the title "National Academy of Medicine" (the McDermott solution). These would be the options that the smaller group would present to the full Board on Medicine when it met in September.[91]

Page and Comroe were less than enthusiastic about the chances for progress. Page described Handler's mood as "flippant" at some times and "fairly harsh" at others. McDermott, he felt, sparred with the Board and represented it "very poorly indeed." Julius Comroe thought that "we are witnessing a last-ditch struggle to retain Walsh McDermott in the leadership (or controlling) position in the affairs of a Board on Medicine or its successor." Page agreed that "most of us have had enough of this cat-and-mouse game, so the field is again wide open, and I am not at all sure that anything profitable will come out of the meeting at the end of the month."[92]

Not much did. Nearly all of the Board members were weary of the long struggle to create a National Academy of Medicine. As one put it, "We have all lost that pleasant little glow . . . because we are haggling over what we are going to call ourselves." "The words are becoming very stale" because "we have not been doing anything for quite some time," London said. To stir things up, the Board agreed that it should return to its program of studies. Even as it tried to get back into the substance of medical policy, however, it kept stumbling over lingering antagonisms. Ivan Bennett mentioned that the Board had failed to advance the torch that the Page group had handed it, and Comroe placed the blame on Walsh McDermott. If the Board had not taken over, Comroe believed, a National Academy of Medicine would have been formed two years earlier.[93]

McDermott told Handler that little of substance had emerged from the meeting, except for a request that the Board meet with the NAS Council, a desire that the Board return to its substantive studies, and a commitment to continue to work with the NAS as the reorganization process progressed. Some Board members, McDermott emphasized, wanted to go further and draw up articles of incorporation for an Academy of Medicine. McDermott thought that support for this action was waning, yet the sentiment persisted. Handler replied that there was little point in the NAS Council meeting with the Board. The Council preferred to wait for the report of the Reorganization Committee, chaired by Franklin Long of Cornell University, before such a meeting.[94]

At this point, neither group knew if it could trust the other. Handler and the NAS Council regarded the creation of a National Academy of Medicine as a real threat to the integrity of the NAS. Ivan Bennett, who was becoming increasingly militant in his quest for a National Academy of Medicine, believed that the NAS had always been "reserved and academic in making studies and proposals," with little "activism in trying to see to it that these proposals are heard in the proper circles and that measures are taken to insure their implementation." Irvine Page thought that there was "little hope of coming to any sensible agreement with NAS without sacrificing the larger objectives of a National Academy of Medicine. I have no enthusiasm whatever to be dominated by Phil Handler and Walsh McDermott, who insist on treating us as second class citizens." The notion of trying to get back to the substantive work of the Board was only a "ploy" to silence Page and his allies.[95]

If so, it worked. For a time in late 1969, the Board concentrated on such substantive issues as the movement to pass national health insurance and the increased use of physician's assistants. A spurt of renewed 1960s optimism attended this phase of the Board's work. It appeared likely that Congress would approve some form of national health insurance and that troop reductions in Vietnam would create a peace dividend that could be spent on social programs. At a time of fears of a shortage of physicians, nurses, and other key health professionals, de-escalation of the war also had the salutary effect of increasing the number of medical doctors and nurses at home. Caught up in the spirit, one Board member talked of opening "up for this nation and our society a whole new vista of progress."[96] Buoyed by this optimistic feeling, the members made a crucial concession to the NAS and agreed to wait for the report of the Long Reorganization Committee before taking action on an independent National Academy of Medicine. "In my judgment," a relieved McDermott told Handler, "the immediate crisis has been weathered." For his part, Handler believed that "our problems are not solved and that we have simply bought time to bring the house in order. But that is all that I had hoped to accomplish at this period."[97]

Some of the fallout from the protracted struggle began to appear in early 1970 in the form of critical articles in the professional press. The standard line was that the Board had failed to live up to its expectations, as demonstrated by the rejection of its plan to investigate medical education. The rebuff of the academy idea by the NAS had "shattered the morale of many board members and staff." As further evidence of the low morale, the press pointed to the fact that both Shannon and his protégé Joseph Murtaugh were planning to leave the National Academy of Sciences. Shannon, who had been acting as an NAS consultant since his 1968 departure from the Academy, took a job with former NAS president Fred Seitz at Rockefeller University. Murtaugh, who had done a heroic amount of staff work for the Board, accepted a position with the Association of American Medical Colleges.[98]

Despite bad press notices, the Board made real progress toward the resolution of its issues in January 1970. For one thing, Ivan Bennett met with the members of the Long committee and came away impressed. Although he had been an outspoken critic of Walsh McDermott, he began to believe that some sort of accommodation with the NAS might be possible. For another, Handler, stressing that the NAS had not vetoed the idea—as opposed to the title—of a National Academy of Medicine, worked at selling the concept of what McDermott called an "Institute of Health Sciences" to Board members.

Page began to give ground. If there could be a broadly based organization that "subtended all of medicine in its broadest sense" and if money could be raised for such an organization, he "would see no reason particularly for starting anything outside" of the Board on Medicine. To be sure, Page remained skeptical, yet he felt he had no alternative but to give the initiative a chance.[99] When he left the January Board meeting, Page told Comroe that he had the strong impression he "would have to play along with it," in part because Robert J. Glaser and Irving London had gone "over to the other side completely." "I suppose in common decency that we have to wait," Page advised.[100]

As Joseph Murtaugh cleared off his desk before leaving the Board on Medicine, he too detected some progress. The Board had decided to turn its energy toward examining the implications of national health insurance for medicine, medical education, and health services. It was the usual broad and diffuse Board inquiry, yet it indicated that at least some members were beginning to think of something other than a National Academy of Medicine. In addition, Dr. Eugene Stead had started a small study of the role of the physician's assistant in medicine. Even better, Murtaugh thought that Page would call off a planning meeting of his "Cleveland group."[101]

In May, the Board on Medicine received a full briefing on the Long committee report. Although the report contained many suggestions about dividing the NAS membership among several subacademies, it also included the idea of creating institutes within the NAS structure. Most of the committee's recommendation required approval of the membership, which could take at least a year. McDermott stressed, however, that the Board could proceed with the creation of an institute immediately. Much like the old Board on Medicine, the new institute would survey the field, generate its own agenda, and undertake its own studies with its own staff. If the Board wished to go this route, "we are in a position to pretty much carve it out the way we want it," said McDermott. Adding his support, London noted that if Page gave his approval and the members approached the task with enthusiasm, the necessary money could be raised to make a go of this institute. McDermott persuaded the Board to put together a proposal for an institute that he could bring to the NAS Council.[102]

The Institute of Medicine

Once again, the Board on Medicine dusted off a proposal and presented it to the Council of the National Academy of Sciences. In this proposal, the Board asked that it be converted to an Institute of Medicine, reporting directly to the NAS Council. It would be "qualitatively similar to the Board but sufficiently expanded . . . to meet the great needs revealed by 2.5 years of Board experience." The Institute would conduct its own studies, provided they were less than a year in duration. Longer studies would be conducted jointly by the NRC and the Institute. As in the previous proposals for an academy, the Institute would have fixed terms of membership and be able to select its own members, most of whom would not be NAS members. At least 25 percent of the members would have degrees in fields other than medicine or biomedical sciences. All members would be selected for their personal attributes, not because they represented a particular constituency. Soon after the Institute started, it would have about 100 members, and ultimately it might grow to 250 members.[103]

Walsh McDermott and a delegation from the Board on Medicine presented this "draft charter" to the NAS Council on June 5, 1970. Yarmolinsky, the principal writer, hastened to reassure Council members that the Charter marked a revision of, but not a departure from, the Board on Medicine. It would not establish a propaganda body, as some Council members feared. On the contrary, it represented the culmination of long years of discussion and established a workable relationship with the NAS. If the Academy rejected this approach, McDermott warned, it would lose "the medical constituency."

The following day, the Council voted to authorize the president of the National Academy of Sciences "to take the necessary steps to create an Institute of Medicine." The Council specified that it would have final authority over the Institute, including the right to review publications and the right to add or delete names of those nominated for membership. It also asked to see a "specific detailed plan describing the organization and operation of the Institute." It regarded this plan as little more than a formality, however, because it authorized Handler to announce the Council decision to the public.[104]

On June 10, 1970, the National Academy of Sciences made a formal announcement that it would create an Institute of Medicine "to address the larger problems of medicine and health care." In a letter to McDermott, Handler cited the design of the health care delivery systems, the role of university medical schools, the mechanisms for funding medical care, and the support and nature of biomedical research as the types of problems the Institute might address. He also requested McDermott to prepare a document containing the bylaws of the organization in time for the NAS Council meeting in August.[105]

Appearing before the Board on Medicine in June, Handler assured the group that the NAS Council had given the Board on Medicine the "broadest kind of hunting license" to establish the new Institute. The head of the Institute would be "someone of stature who would command a commensurate salary." The Board envisioned such a person to be about 45 years old. "If you find the best man for it, we are just going to be delighted," Handler said, using the conventional sexist language of the era. He added that he did not foresee any problems with NAS Council approval.[106]

A draft of the constitution and bylaws received the Board's approval on July 22. The next step was for Irving London, Adam Yarmolinsky, Walsh McDermott, and James Shannon to present these documents to the NAS Council on August 24. In the meantime, Handler offered his suggestions. Among other things, he changed the title of the president of the IOM to the director, and the name "executive council" to "executive committee." In making these suggestions, Handler hoped to avoid confusion between the IOM and the NAS itself.

The August 24 meeting of the NAS Council went smoothly. On nearly all matters of dispute between the Board and the NAS, the Council yielded. The Council wanted to restrict the total number of members, but McDermott assured them that the Institute's work required more than 200 people. The Council agreed that the IOM should have a president, not a director. Both sides concurred in the provision that all members of the Medical Sciences Section of the National Academy of Sciences would receive an invitation to join the Institute.

On September 9, 1970, Walsh McDermott wrote that he, Adam Yarmolinsky, and Irving London were all quite pleased with the final version of the Institute of Medicine's Charter and Bylaws. The Charter contained a ringing statement that reflected McDermott's point of view: "The problems posed in provision of health services are so large, complex and important as to require, for their solution, the concern and competences not only of medicine but also of other disciplines and professions."[107] In the terms of the earlier debate with Irvine Page, the Institute would speak about, not for, medicine.

Irvine Page retreated as gracefully as he could. In an editorial that appeared in October, he stated that he had once believed that only a freestanding National Academy of Medicine would do. Recently, however, he had been persuaded that the "N.A.S sincerely wishes to fulfill those needs through an Institute of Medicine. The immediate plans have taken more than two years to evolve, and the institute deserves a fair trial." Robert J. Glaser called Page's statement "excellent. . . . I believe we are now well on our way to creating the kind of organization you and others envisioned from the start."

Although Glaser warned that much would depend on securing financial backing, the initial signals from the foundation community were encouraging. Independently of Glaser, Page learned that foundation money for the sort of national academy he had wanted to form was no longer obtainable. As Page put it, the Carnegie Foundation and Commonwealth Fund planned to let the Institute "have a try and fail before big money will be available for NAM."[108]

Conclusion

Between 1967 and 1970, the Board on Medicine spent most of its time trying to reconcile the conflicting points of view of its members and negotiating with the National Academy of Sciences. As a consequence, it completed few studies and gained fame only for a short statement that it issued on heart transplants. An initial round of enthusiasm for the venture went largely unexploited, because the members of the Board could never quite agree on its final form. An almost superhuman form of patience was required on Walsh McDermott's part to bring the process to conclusion.

In the end, however, the Institute of Medicine reflected McDermott's vision as much as anyone else's. When Irvine Page wrote the 1964 editorial launching the process that ultimately led to the Institute of Medicine, he did not envision that the result would be an organization in which a quarter of the membership would be composed of people from outside the health sciences. Nor did he foresee an entity concerned not only with the problems of medical research but also with the larger social problems surrounding health. The Institute of Medicine was created at a time when health policy experts and even some of those in the realm of heath sciences research, such as McDermott, emphasized the social aspects of medical problems and people thought these problems were amenable to solution. The optimism of the era was stamped on the Institute of Medicine's Charter and Bylaws and on its initial agenda.

So were the organizational tensions that had developed in the National Academy of Sciences. These tensions created ambiguities that would continue to plague the Institute. Exactly how much autonomy did the Institute have? How did the Institute relate to the rest of the Academy structure? Few people wished to probe these questions too deeply for fear of upsetting the delicate set of understandings on which the IOM rested.

On December 17, 1970, Philip Handler appeared before the Board on Medicine, functioning as the Executive Committee of the Institute of Medicine, and announced that Dr. Robert J. Glaser would serve as acting president of the Institute. He would volunteer his time and keep his regular job. Until recently the dean of Stanford's School of Medicine, Dr. Glaser currently worked as a vice president of the Commonwealth Fund in New York, a position previously held by Board member Colin MacLeod. Four days later, the NAS issued a press release announcing the "formal activation of the Institute of Medicine."[109] The Institute of Medicine was launched.

Notes

1. Irvine H. Page, "Needed—A National Academy of Medicine," *Modern Medicine,* July 20, 1964, pp. 77–79; Wilbur Cohen, "Health and Security for the American People," reprinted in U.S. Senate, Committee on Finance, *Nomination of Wilbur J. Cohen*, March 22 and 23, 1961. The description of the Association of American Medical Colleges comes from Robert Ebert, "The Changing Role of the Physician," in Carl J. Schramm, ed., *Health Care and Its Costs* (New York: W.W. Norton and Company, 1987), p. 163. For a historical overview of the many organizations concerned with health policy, see Paul Starr, *The Social Transformation of American Medicine* (New York: Basic Books, 1982).

2. Lawrence K. Altman, "Dr. Irvine H. Page Is Dead at 90: Pioneered Hypertension Research," *New York Times*, June 12, 1991; Curriculum Vitae of Irvine H. Page, M.D., Box 8, Irvine Page Papers, National Library of Medicine, Bethesda, Md.

3. Paul Starr, *The Social Transformation of American Medicine*, p. 347; Stephen P. Strickland, *Politics, Science, and Dread Disease: A Short History of United States Medical Research Policy* (Cambridge, Mass.: Harvard University Press, 1972); Daniel M. Fox, *Health Policies, Health Politics: The British and American Experience, 1911–1965* (Princeton, N.J.: Princeton University Press, 1986).

4. Irvine H. Page, "More on a National Academy of Medicine," *Modern Medicine*, March 15, 1965, pp. 89–90; materials in File IOM: 1965, Box 1, Page Papers.

5. Irvine Page to Dr. Karl Folkner, President, Stanford Research Institute, April 6, 1964, File IOM: 1965, Page Papers.

6. Irvine Page, "Membership in a National Academy of Medicine," and "Replies to July 1965 Query re: Membership in a N.A.M.," both in Container 1, File IOM: 1965, Page Papers.

7. Edward D. Berkowitz, *America's Welfare State: From Roosevelt to Reagan* (Baltimore, Md.: Johns Hopkins University Press, 1991), p. 178; Stephen P. Strickland, *Politics, Science, and Dread Disease*, p. 207.

8. "News in Brief," *Science*, September 23, 1966, p. 1508.

9. Irvine Page to Robert Aldrich, September 6, 1966, and "Agenda of First Organizational Meeting," January 17, 1966, both in Page Papers.

10. Irvine Page, "Is There a Need?" Notes for first organizational meeting, Page Papers.

11. Donald S. Fredrickson, "James Augustine Shannon," *Proceedings of the American Philosophical Society*, 140 (March 1966), pp. 107–114; Stephen P. Strickland, *Politics, Science, and Dread Disease*, pp. 100–101.

12. "Minutes of Organizational Meeting of National Academy of Medicine," January 7, 1967, Board on Medicine Files, National Academy of Sciences (NAS) Archives.

13. "Minutes of Second Organizational Meeting for a National Academy of Medicine," March 7, 1967, Board on Medicine Files, NAS Archives.

14. Ross Harrison to George H. Whipple, September 6, 1941, and E. B. Wilson to Harrison, September 12, 1941, Box 80, Ross Harrison Papers, Yale University.

15. *Report of the National Academy of Sciences, National Research Council*, Fiscal Year 1945–46 (Washington, D.C.: U.S. Government Printing Office, 1947), p. 52.

16. See, for example, National Academy of Sciences, National Academy of Engineering, National Research Council, *Annual Report Fiscal Year 1967–68* (Washington, D.C.: U.S. Government Printing Office, 1970).

17. Frederick K. Seitz Memoir, p. 289; J. F. A. McManus, M.D., "Why a National Academy of Medicine?" *Medical Science*, April 1967, in Agenda Book, Background Materials on Further Institutional Forms for the Board on Medicine, Board on Medicine Files.

18. Council Meeting, Executive Session, April 22–23, 1967, and Business Session, April 25, 1967, both in Board on Medicine: General 1967 file, NAS Records.

19. Council Meeting, June 3 and 4, 1967, and Governing Board Meeting, June 4, 1967, both in Board on Medicine: General 1967 file, NAS Records.

20. Betty Libby to Robert Aldrich, June 15, 1967, and Irvine Page to Philip Lee, June 23, 1967, both in Page Papers.

21. Walsh McDermott to Irvine Page, October 5, 1966, File IOM: 1966, Page Papers.

22. Paul B. Beeson, "Walsh McDermott, October 24, 1909–October 17, 1981," NAS *Biographical Memoirs*, pp. 283–299, enclosed in Beeson to Samuel Thier, May 10, 1990, Wallace Waterfall Materials, National Academy of Sciences.

23. Frederick Seitz to Dr. Walsh McDermott, June 28, 1967, and McDermott to Seitz, July 22, 1967, NAS, Committee and Boards, Board on Medicine, General, NAS Records, NAS Archives.

24. Minutes of Meeting, June 28, 1967, Board on Medicine Meetings, Minutes, 1967–1968, Board on Medicine Files, NAS Records, NAS Archives.

25. Irvine H. Page to Dr. Robert Aldrich et al., September 13, 1967, Container I, File IOM: August–December 1967, Page Papers; Diary Note, September 14, 1967, Meeting of Organizing Group, September 12, 1967, NAS Records.

26. Minutes of Council Meeting, September 30, 1967, NAS Records; Walsh McDermott to Frederick Seitz, September 18, 1967, NAS Records; Daniel Bell to Seitz, October 3, 1967; Kermit Gordon to Seitz, October 26, 1967; Milton Friedman to Seitz, October 13, 1967; and James Tobin to Seitz, October 5, 1967, all in Board on Medicine: Membership, NAS Records; Seitz to Rashi Fein, October 31, 1967, and Fein to Seitz, November 6, 1967, both in Rashi Fein Papers (privately held); Interview with Rashi Fein, August 2, 1997, Lake George, New York.

27. Evert Clark, "Medical Board Set Up to Speed Benefits of Research to Public," *New York Times*, November 14, 1967, p. 1. Other media coverage included articles in the *Journal of the American Medical Association*, December 4, 1967; *Medical World News*, November 24, 1967; *U.S. Medicine*, December 15, 1967, *Medical Tribune*, November 27, 1967; and *Science*, 158 (1967), p. 891. Each of these stories is carefully collected in the Page Papers.

28. Irvine Page to Colin MacLeod, July 2, 1967; Page to Fred Seitz, October 3, 1967; Page to Stuart M. Sessoms, December 4, 1967; Page to James A. Shannon, December 8, 1967; Page to Julius Comroe, January 2, 1968; and Page to MacLeod, January 2, 1968, all in Page Papers.

29. "First Meeting of the Board on Medicine, National Academy of Sciences, November 15, 1967: Summary of Discussion" (prepared by Joseph Murtaugh), Board on Medicine Meetings, Minutes, 1967–1968, Board on Medicine Files.

30. "Summary Minutes of the Second Meeting of the Board on Medicine, Washington, D.C.," December 14 and 15, 1967, File Board on Medicine Meetings: Minutes 1967–1968, Board on Medicine Files, NAS Records, NAS Archives.

31. Walsh McDermott to Dr. Julius H. Comroe, Jr., January 16, 1968, File Functional Statement: Proposed, Board on Medicine Files.

32. Irvine H. Page to Drs. R. A. Aldrich et al., December 18, 1967, File Board on Medicine: Meeting: December 14–15, 1967, Board on Medicine Files.

33. Irvine H. Page to Drs. R. A. Aldrich et al., February 6, 1968, Container 1, Page Papers.

34. Minutes of Board on Medicine Meeting, February 2, 1968, Board on Medicine Files.

35. Walsh McDermott to Board on Medicine, February 20, 1968, Board on Medicine Files.

36. "Board on Medicine Recommends Criteria for Heart Transplants," press statement for release on February 28, 1968, File Draft Annual Report, August 1969, Board on Medicine Files.

37. Evert Clark, "Guidelines Urged for Transplants," *New York Times*, February 27, 1968.

38. Rudy Abramson, "Don't Try Heart Transplants too Soon, Scientists Warn," *Atlanta Journal*, February 29, 1968.

39. "Appraising Heart Transplants," *Saturday Review*, April 6, 1968, pp. 59 ff.

40. Alberto Z. Romualdez, M.D., Secretary General, World Medical Association, to Walsh McDermott, February 2, 1968, Heart Transplant

Material, Board on Medicine Records; *Congressional Record*, March 5, 1968, pp. S 22212–22213; Joseph A. Califano, Jr., to McDermott, March 5, 1968; Wilbur Cohen to McDermott, March 11, 1968; and Joseph Murtaugh to McDermott, April 21, 1968, all in File Transplant Statement: Correspondence, Board on Medicine Records.

41. Dr. Francis Moore to Walsh McDermott, March 14, 1968, File Transplant Statement: Correspondence, Board on Medicine Records.

42. Walsh McDermott to Dr. Francis D. Moore, April 3, 1968, File Transplant Statement: Correspondence, Board on Medicine Records.

43. Walsh McDermott to Dr. Henry K. Beecher, Department of Anaesthesia, Harvard Medical School, April 27, 1965, File Transplant Statement: Correspondence, Board on Medicine Records.

44. Irvine Page to Walsh McDermott, March 18, 1969, File Transplant Statement: Correspondence, Board on Medicine Records.

45. Quigg Newton, Commonwealth Fund, to Fred Seitz, May 9, 1968, Board on Medicine Records.

46. James Shannon to Fred Seitz, May 23, 1968, Committee and Boards: Board on Medicine, General, 1968, NAS Records.

47. Joseph Murtaugh to the record, Meeting at the Commonwealth Fund, June 10, 1968, Committees and Boards, 1968, Board on Medicine, Panels, Further Institutional Forms Ad Hoc, NAS Records.

48. See "Source of Funds, Board on Medicine, 1968–1969," Board on Medicine Records.

49. "Summary Report of Meetings April 10 and 11, at Stone House, Bethesda, Maryland," April 15, 1968, and Joseph Murtaugh to John S. Coleman, April 22, 1968, File Meeting April 10–11, 1968, Agenda, Attendance, Draft Minutes, both in Board on Medicine Files;

50. Task Order 49, Medical Education for the Future, Board on Medicine Records.

51. "Health and the Poor" (Proposal for a Study), June 10, 1968, Study of Health and the Disadvantaged File, Board on Medicine Files.

52. "Summary Report of the Meeting of the Panel on Health and the Disadvantaged," September 11, 1968, Study of Health and the Disadvantaged File, Board on Medicine Records.

53. Thomas Francis, Jr., M.D., Professor and Chairman, Department of Epidemiology, University of Michigan, to Fred Seitz, January 5, 1968, Committee and Boards: Board on Medicine: General, 1968, NAS Records,

54. Internal Health, Education, and Welfare memorandum from Jane E. Fullarton to the Record, March 25, 1968, included in Joseph Murtaugh to Members, Board on Medicine, April 4, 1968, Drawer 1, Board on Medicine Files; Julius Comroe to McDermott, April 8, 1968, File Board on Medicine Functional Statement: Proposal, Board on Medicine Files.

55. Julius Comroe to Joseph Murtaugh, March 14, 1968, and Irvine Page to Comroe, March 19, 1968, File NAS-Board on Medicine Correspondence, March 1968, Page Papers.

56. Irvine Page, "Needed—A National Academy of Medicine," July 3, 1968, in Joseph Murtaugh to Members, Board on Medicine, July 16, 1968,

NAS Committees and Boards (1968), Board on Medicine: Panels, Further Institutional Forms Ad Hoc, NAS Records.

57. Julius Comroe form letter, July 8, 1968, in Meeting July 22, 1968: Agenda, Attendance File, Board on Medicine Records; Comroe to Board on Medicine, July 26, 1968, in Agenda Book, Background Materials on Further Institutional Forms for the Board on Medicine, Board on Medicine Records; File July 22, 1968, Meeting Transcript, p. 42, Board on Medicine Records.

58. Julius Comroe to Irvine Page, August 16, 1968, File NAS-Board on Medicine, July–September 1968, Page Papers. The members of the panel were Irving London, Julius Comroe, Robert J. Glaser, Henry Riecken, Walter Rosenblith, and Dwight Wilbur, with James Shannon listed as a consultant.

59. Board on Medicine Meeting Transcript, September 25, 1968, p. 164, Board on Medicine Files, "Notes on the Meeting of the Ad Hoc Panel of Further Institutional Forms for the Board on Medicine," September 27, 1968, 1968, NAS Records.

60. Joseph Murtaugh to Fred Seitz, October 8, 1968, NAS Records.

61. Irving S. Wright to Irvine Page, November 13, 1968, Container 2, File NAS-Board on Medicine Correspondence, November–December 1968, Page Papers.

62. Carleton B. Chapman, M.D., Dean, Dartmouth Medical School, to Irvine Page, December 18, 1968; Page to Chapman, December 27, 1968; and Page to Francis D. Moore, December 26, 1968, all in File NAS-Board on Medicine Correspondence, November–December 1968, Page Papers.

63. Irvine Page to Irving London, November 14, 1968, NAS Committees and Board, Future Institutional Forms, NAS Records.

64. Irvine Page, "The Two-Concept Idea," Background Materials for Further Institutional Forms, Board on Medicine Files.

65. "Notes on the Meeting in Boston, November 18, 1968, of the Ad Hoc Panel on Further Institutional Forms, Agenda Book, Background Materials for Further Institutional Forms, Board on Medicine Files.

66. Transcript of Board on Medicine Meeting, November 20–21, 1968, Board on Medicine Records. See particularly pp. 133–134.

67. Irvine Page to Francis Moore, December 26, 1968, and Dwight Wilbur to Page, January 3, 1969, both in File NAS-Board on Medicine Correspondence, January–February 1969, Page Papers.

68. Joseph S. Murtaugh to Members, Board on Medicine, January 9, 1969, in Drawer One, Board on Medicine Records; NAS Press Release, January 17, 1969, Drawer One, Board on Medicine Records.

69. "Summary Notes on the Meeting of January 20, 1969," January 27, 1969, and Robert A. Kevan to Joseph Murtaugh, January 28, 1969, both in Study on Biomedical Education for the Future File, Board on Medicine Records.

70. Robert Glaser to Irvine Page, February 13, 1969, and Page to Glaser, February 20, 1969, both in File NAS-Board on Medicine Correspondence January–February 1969, Page Papers.

71. Irving M. London to Walsh McDermott, March 7, 1969, NAS Records.

72. "A Proposal for a National Academy of Medicine," February 28, 1969, Committee and Boards, Panel: Further Institutional Forms, Ad Hoc, General, 1969, NAS Records.

73. Irvine Page to Dwight Wilbur, March 3, 1969, File NAS-Board on Medicine Correspondence March–April 1969, Page Papers.

74. Transcript of Board on Medicine Meeting, March 12, 1969, pp. 113, 126–127, 256, Board on Medicine Records; Walsh McDermott to Irvine Page, March 21, 1969, and Julius Comroe to Page, March 17, 1969, both in File NAS-Board on Medicine Correspondence March–April 1969, Page Papers.

75. Walsh McDermott to Fred Seitz, March 24, 1969, NAS Council, Executive Committee Meeting, March 29, 1969, NAS Records; McDermott to Board on Medicine, May 22, 1969, NAS Records; McDermott to Irving London, April 15, 1969, NAS Records; Minutes of NAS Council, April 26–27, 1969, NAS Records.

76. Transcript of Board on Medicine Meeting, May 7–8, 1969, pp. 53, 105–108, Board on Medicine Files.

77. Joseph S. Murtaugh to Members, Board on Medicine, "Summary Report on Board on Medicine Meeting, May 7–8, 1969," McDermott Files, Board on Medicine Records; McDermott to the President and Members of the Council of the National Academy of Sciences, May 29, 1969, Agenda Book, June 25–26, 1969, Meeting, Board on Medicine Records.

78. Ronald Kotulak, "Plan Super Agency to Advise on Health," *Chicago Tribune*, April 25, 1969; David Rhinelander, "Health Expert Wants Medicine Academy," *Hartford Courant*, May 4, 1969; and "Scheme for Medical Academy," *Nature*, 223 (August 9, 1969).

79. Minutes of the Council Meeting, June 7, 1969, NAS Records; Joseph Murtaugh to McDermott, Glaser, London, Yarmolinsky, and Shannon, June 12, 1969, NAS Records.

80. Council Meeting, June 7, 1969, Executive Session, Restricted, NAS Records.

81. Philip Handler to the Council, National Academy of Sciences, June 17, 1990, McDermott Files, Board on Medicine Records.

82. Joseph Murtaugh to Members, Board on Medicine, June 11, 1969, NAS Records.

83. NAS Executive Committee Meeting, July 19, 1969, NAS Records.

84. Joseph Murtaugh, "Notes on the Meeting of the Delegation of the Board on Medicine with the Executive Committee of the NAS Council, July 20, 1969, at Woods Hole, Massachusetts," July 28, 1969, NAS Records.

85. Philip Handler to Walsh McDermott, July 29, 1969, NAS Records.

86. Transcript of Board on Medicine Meeting, July 30, 1969, Board on Medicine Records.

87. Irvine Page to James Shannon, August 1, 1969, and Page to Cecil Andrus, August 6, 1969, both in File NAS-Board on Medicine Correspondence July–August 1969, Page Papers.

88. Robert Q. Marston to Ivan Bennett, June 20, 1969, Meeting File for September 29–30, 1969, Board on Medicine Records; Robert A. Kevan to C. H.

William Ruhe, August 7, 1969, Reorganization File for the Study on Biomedical Education for the Future, Board on Medicine Records.

89. Philip Handler to Philip H. Abelson et al., August 1, 1969, "Background Materials," September 9–10, 1969, Meeting of the Ad Hoc Committee on Further Institutional Forms of the Board on Medicine, Board on Medicine Records; Telegram from Walsh McDermott to Philip Handler, August 5, 1969, National Academy of Medicine Proposal File, 1969, Board on Medicine Records; Handler to Joseph Murtaugh, August 5, 1969, McDermott Files, Board on Medicine Records; Handler, memoranda for file, August 12 and August 13, 1969, NAS Records.

90. Philip Handler to Walsh McDermott, August 12, 1969, Board on Medicine: Meetings, Resolution to Create National Academy of Medicine, July 1969 File, NAS Records; McDermott to Handler, n.d., NAS Records.

91. "Meeting of the Ad Hoc Panel on Further Institutional Forms, New York City, September 9–10, 1969," Board on Medicine Records.

92. Irvine Page to Julius Comroe, September 11, 1969, Container 2, File NAS-Board on Medicine Correspondence September–December 1969, Page Papers; Comroe to Page, August 25, 1969, NAS-Board on Medicine Correspondence July–August 1969, Page Papers.

93. Transcript of Board on Medicine Meeting, September 29–30, 1969, Board on Medicine Records.

94. Walsh McDermott to Philip Handler, October 6, 1969, Bennett Committee File, Board on Medicine Records; Handler to McDermott, September 8, 1969, McDermott Files, Board on Medicine Records; NAS Council Minutes, September 27, 1969, NAS Records.

95. Ivan Bennett to Irvine Page, October 16, 1969, File NAS-Board on Medicine Correspondence September–December 1969, Page Papers; Page to Bennett, November 4, 1969, File NAS-Board on Medicine Correspondence November–December 1969, Page Papers.

96. "Summary Report of the Meeting of the Board on Medicine," November 12, 1969, Board on Medicine Records.

97. Walsh McDermott to Philip Handler, November 13, 1969, and Handler to McDermott, November 25, 1969, NAS Records.

98. "NAS Spikes Proposal for Academy of Medicine," *Biomedical News*, January 1970; "Ivory Tower of Medicine: Is 'No' Final?" *Medical World News*, January 2, 1970, both in Institute of Medicine Proposed, 1970 File, NAS Records; *Research Notes*, January 7, 1970, Institute of Medicine—General File, NAS Records.

99. Transcript of Board on Medicine Meeting, January 21–22, 1970, Board on Medicine Records.

100. Irvine Page to Julius Comroe, January 22, 1970, and Page to Gerald S. Berenson, M.D., February 19, 1970, both in File NAS-Board on Medicine Correspondence, January–April 1970, Page Papers.

101. Joseph Murtaugh to John Coleman, January 30, 1970, Loose File, Board on Medicine Records.

102. Transcript of Board on Medicine Meeting, May 13–14, 1970, Board on Medicine Records.

103. "Proposal to Council of NAS by Board on Medicine," June 5, 1970, NAS Records.

104. Minutes of the Council Meeting, June 5–6, 1970, NAS Records.

105. NAS Press Release, June 10, 1970, NAS Records; Philip Handler to Walsh McDermott, June 10, 1970, NAS Records.

106. Transcript of Board on Medicine Meeting, June 24, 1970, Board on Medicine Records; "Brief Minutes," June 29, 1970, McDermott Files, Board on Medicine Records.

107. Institute of Medicine, Charter and Bylaws, 1970, NAS Records.

108. Douglas Bond, M.D., President, Grant Foundation, to Irvine Page, July 21, 1970, and Robert Glaser to Page, August 10, 1970, both in File NAS-Board on Medicine July–September 1970, Page Papers; Irvine Page, "The Institute of Medicine of the National Academy of Sciences," *Modern Medicine*, October 5, 1970, pp. 95–97.

109. "Dr. Robert J. Glaser Named Acting President of Institute of Medicine," NAS Press Release, December 21, 1970, NAS Records; Institute of Medicine, Meeting of Executive Committee, December 17, 1970, NAS Records.

2

The Institute of Medicine
Begins Operations

Early in 1971, Philip Handler, the president of the National Academy of Sciences (NAS), wrote one of his correspondents that "the Institute of Medicine [IOM] is, at the moment, largely a paper organization, but the members of our Board on Medicine are busily engaged in fleshing it out."[1] Four and a half years later, after two IOM presidents had come and gone, Roger Bulger, a key IOM staff member, noted that it had never "been our expectation that we would become a household word across the United States of America. We have never felt that we would become, at least in the first years, an opinion maker at the level of the individual citizen."[2]

As Bulger indicated, the Institute of Medicine attracted little national attention during its first five years of operations. This was because its work did not directly engage the questions that animated the public, such as the passage of national health insurance, nor did its leaders seek to become political celebrities who testified often before Congress. Instead, the IOM maintained a low profile and focused on relatively technical, but nonetheless important, questions related to health policy and conduct.

The new organization faced internal pressures that mitigated against its undertaking more ambitious projects. Members expected to shape the organization's agenda and participate in its activities, yet the IOM depended heavily on Washington-based staff, attuned to the nuances of public policy and accustomed to taking the lead on research projects, to guide its work. The action-oriented IOM formed part of the National Academy of Sciences and had to adapt to its culture. This meant an insistence on intellectual rigor attained by close peer review of Institute reports, combined with a relatively passive attitude toward obtaining grants and contracts. The National Academy of Science's leaders believed that Washington should come to them, an outlook that the Institute of Medicine, a younger, less established organization with a direct dependence on external funds,

50

could ill afford. These differences in institutional outlook between the NAS and the IOM often took the form of personal confrontations between IOM presidents John Hogness and Donald Fredrickson and NAS President Philip Handler.

Even as Hogness and Fredrickson dealt with these constant pressures, they faced the more mundane task of running the IOM on a daily basis. The invention of routines that would guide the new organization through its first five years fell to them. Much of their work centered on the task of recruitment. They had to hire staff members who would set the tone for the Institute's subsequent development, select members who would initiate key committees such as the Program Committee, and make vital contacts with the foundation officers and public officials whose decisions on funding grant proposals held the key to the Institute's very survival.

Neither of these presidents served for a long time. John Hogness spent three years in office before leaving to become president of the University of Washington; Donald Fredrickson stayed at the IOM for less than a year, much of it spent in distracting and ultimately successful negotiations with federal officials over whether he would become the head of the National Institutes of Health (NIH). The turnover made the difficult endeavor of starting the IOM that much harder. Because Hogness and particularly Fredrickson left so quickly, the organization failed to establish a sense of continuity. Those in the foundation and health policy communities in positions to fund the IOM and increase its visibility found it difficult to gain a "fix" on the organization's identity. Nonetheless, Hogness and Fredrickson both made enduring impressions on the Institute of Medicine's history. Under their leadership in the years between 1971 and 1975, the IOM perfected its form of governance through establishment of the IOM Council and experimented with different ways of influencing health policy. IOM leaders came to realize that the organization possessed the means of convening the nation's leading experts to consider important issues in health policy. As a result, the IOM influenced President Nixon's war on cancer and established an important methodology for estimating the costs of medical education. By the end of the period, the IOM, if not a household name, had become known to the Washington health policy community as an organization with the potential to serve as a useful and influential source of impartial advice.

Recruiting Members

After Philip Handler declared the IOM open for business in December 1970, the first task was very basic: the members of the old Board on Medicine, now reconstituted as the IOM Executive Committee, had to pick the initial members of the new Institute of Medicine. This task, like many others in the period between the end of 1970 and the spring of 1971, fell on acting President Robert J. Glaser, a witty man and an accomplished physician, who proved quite adept at the delicate task of getting the organization started. He had been present at the first, tentative NAS meetings that had given birth to the Board on Medicine, and he was a veteran of the long discussions that had led to the Institute of Medicine. During the often contentious sessions that followed, he managed to retain the goodwill of Walsh McDermott, Irvine Page, and Julius Comroe.

Like McDermott, Page, and Comroe, Glaser came from the world of academic medicine. Raised in St. Louis, he went to Harvard at a time when that school was just beginning to accept students from the Midwest and West who lacked a fancy prep school background. Graduating from a Harvard class that also contained John F. Kennedy, Glaser continued his studies at Harvard Medical School. This enabled him to embark on a career that included a series of distinguished academic appointments. At Washington University, he headed the Division of Immunology. In 1957, he moved to the University of Colorado, where he became one of the nation's youngest medical school deans. Six years later, he returned to Harvard, where he held the university's first chair in social medicine. Then, in 1965, he arrived at Stanford to become dean of the medical school. Five years later, Glaser accepted a job as vice president of the Commonwealth Fund, which enabled him to make occasional trips from New York to Washington to handle IOM business on a part-time basis.[3]

Glaser chaired what was known as the Initial Membership Committee. Ivan Bennett, Walsh McDermott, and Eugene Stead also served on the committee that began its work even before the formal creation of the Institute and made many of the important decisions about the terms of IOM membership. Discussions centered on the appropriate age, proper geographic distribution, and right mix of medical specialties, other health professionals, and nonmedical fields. Although the process went smoothly, disagreements with the Executive Committee (the old Board on Medicine) arose over how to treat people who were approaching age 60. The Membership Committee wanted to exclude them; the Executive Committee disagreed, hoping to attract as much talent to the Institute as possible.[4] As a compromise, Glaser's committee decided to apply an age limit of 64. This meant that the IOM would not necessarily be an organization of young men, even though the Charter called for people to become senior members when they reached age 66. Nor would the IOM necessarily be an organization with many active members, although it did work out this way and willingness to work was defined as one of the conditions of membership. The initial Membership Committee decided not to consider whether a person was already overburdened with other commitments before asking him or her to become a member; those asked to join were allowed to make up their own minds about whether they could find time to serve.[5]

Having set the basic ground rules, the Membership Committee proceeded to generate a list that contained nearly twice as many people as necessary to bring the total membership to 100. Board on Medicine and NAS members who belonged to the Section on Medicine received automatic invitations. This left 77 positions to fill. In January 1971, the committee nominated 48 people for membership from a list of 85, and in March the committee selected the final 29 nominees. In contrast to later years when the Institute of Medicine would hold a formal membership election, in 1971 the nominees were selected by a committee, approved by the Executive Committee and Philip Handler, and then asked to join. Glaser presented the final list to Handler in April. Letters to newly selected members went out toward the end of May 1971, and the NAS made the formal announcement in June.[6]

The National Academy of Sciences protected the Institute of Medicine from the more overt forms of lobbying that accompanied the membership process. Ever since the formation of an Institute of Medicine had become a real possibility, Philip Handler had received a stream of letters from organizations arguing that they deserved representation in the IOM. Psychiatrists, pharmacists, toxicologists, veterinarians, dieticians, and rehabilitation doctors all wrote to Handler. To his credit, he kept these letters to himself and did not pressure the IOM to accept anyone merely as a representative of a particular profession or medical specialty. He preferred that merit predominate.[7]

To be sure, the IOM Executive Committee and the NAS Council did not agree completely on the criteria for membership. Although the NAS Council approved the list, some Council members complained that key scientific fields, such as demography and epidemiology, did not receive enough representation. The IOM, for its part, worried about the lack of practicing physicians, who constituted only 10 percent of the initial membership. In a variation of a discussion that had gone on since 1967, Glaser told Handler that practicing physicians were "essential for the Institute to carry out its obligations," even though their accomplishments could not be measured by publications or membership in prestigious professional societies. Glaser even admitted that in the absence of these academic criteria, the selection process for practicing physicians was less rigorous than for other members. Identifying physicians engaged in the private practice of medicine who had national reputations proved a difficult task. Some of the physicians who were chosen, such as one who practiced near the NAS's unofficial summer headquarters in Woods Hole, Massachusetts, owed their selection to the fact that key IOM and NAS officials knew them. In addition to the concern over finding enough private physicians with sufficient distinction to merit IOM membership, Glaser worried about the geographic tilt of the final list. A disproportionate number of members came from the Northeast, mid-Atlantic, and West Coast regions—a taint of elitism that bothered the IOM far more than it did the NAS.[8]

Charter Members of the Institute of Medicine

Paul B. Beeson, M.D.	Walsh McDermott, M.D.*
Ivan L. Bennett, Jr., M.D.*	Carl V. Moore, M.D.*
Charles G. Child III, M.D. *	Samuel M. Nabrit, Ph.D.
Julius H. Comroe, Jr., M.D.*	Irvine H. Page, M.D.*
Jerome W. Conn, M.D.*	Henry W. Riecken, Ph.D.
Rashi Fein, Ph.D.	Walter A. Rosenblith, Ing.Rad.
Robert J. Glaser, M.D.	Ernest W. Saward, M.D.*
Robert A. Good, M.D., Ph.D.	James A. Shannon, M.D., Ph.D.*
Leon O. Jacobson, M.D. *	Thomas H. Weller, M.D.
Henry G. Kunkel, M.D.*	Dwight L. Wilbur, M.D. *
Lucile P. Leone, M.A.	Bryan Williams, M.D.
Irving M. London, M.D.	W. Barry Wood, Jr., M.D.*
Colin M. MacLeod, M.D.*	Adam Yarmolinsky, LL.B.
Maclyn McCarty, M.D.	Alonzo S. Yerby, M.D.*

*Deceased

The group did better in selecting members from a broad array of fields. Glaser identified 19 different fields in the initial membership group, including administration, basic sciences, engineering, community medicine, dentistry, nursing, and nutrition.[9] It was, by any sort of measure, an impressive group. It contained two future IOM presidents, Donald Fredrickson and David Hamburg. It featured people who practiced medicine in very different settings: a family practitioner from Hampton Highlands, Maine, and the dean of the Harvard Medical School; the chairman of the Johns Hopkins Department of Pediatrics and a nurse–midwife from New York Downstate Medical Center. The physician-in-chief at Beth Israel Hospital, the general director of Massachusetts General Hospital, and the head of Blue Cross–Blue Shield all made their way to the Institute of Medicine. Only four people, who were not directly involved in health policy and had corporations or other large organizations to run, turned down the IOM's offer of membership.[10]

John Hogness

At the same time that Robert J. Glaser orchestrated the talent hunt for IOM members, he also coordinated the search for the first permanent IOM president. In December 1970, Glaser himself was on a short list of eight candidates, but he made it known that he was not interested; he preferred to launch the organization and do his work at the Commonwealth Fund. Like Glaser, all of those on the list were white, male medical doctors who worked in an academic or a research setting; indeed, everyone who headed the IOM in its first quarter century fit this description. Because each person on the list held a prestigious position such as chief of medicine, chairman of a medical school department, or medical school dean, Glaser realized it would require salesmanship to interest one of these individuals in the IOM job.

On January 20, 1971, Glaser reported to Philip Handler on a meeting with John Hogness, former dean of the Medical School at the University of Washington and current director of the university's Health Sciences Center. "I got the feeling," Glaser noted, "that John is extremely interested and that the chances are very good that he will be interested in taking it on." Two months later, Handler made a formal offer to Hogness to serve a five-year term beginning on July 1, 1971. On March 30, 1971, the Academy announced Hogness's selection. "With the appointment of Dr. Hogness, the Institute of Medicine becomes a reality," said Handler.[11]

On the same day, the *New York Times* made John Rusten Hogness its "Man in the News," indicating a high level of interest in the IOM and its affairs. The piece featured a picture of Hogness, stethoscope around his neck, intently performing a medical examination. His hair slicked back, Hogness looked a bit like the television reporter Mike Wallace. The head and shoulders shot gave no indication of Hogness's height—6 feet, 4 inches—"the size of a tackle," according to the *Times*, nor did the grainy black-and-white photo capture the blondish cast of his hair. Physically imposing, Hogness related well to people and used his self-deprecating sense of humor to soften what the *Times* described as "his vigorous and innovative way of doing things."[12]

Although Hogness later wondered what factors played a role in his selection as IOM president, he was a natural enough choice. His selection helped to bridge some of the gaps between the McDermott and Page factions of the Board on Medicine. Although he came from an academic background at the University of Washington, he had maintained a private practice in Seattle for most of the 1950s. Leaving private practice in 1959, he plunged into the administration of the academic medical center at the University of Washington, serving first as medical director of the newly opened university teaching hospital, next as dean of the School of Medicine, and finally as executive vice president of the University of Washington at a time when the disruptions caused by the Vietnam War made it a trying proposition. He persevered and developed a reputation as an excellent university administrator, one who received regular offers from large research-oriented universities to serve as president. Not only had he been in private practice, he had served as secretary–treasurer of the local medical society. Not only had he done research in endocrinology, he had also chaired the Board of Health Sciences at the University of Washington and come to know the members of other health professions. He therefore had connections with both the nationally oriented academic and the locally oriented private practice sides of medicine and with its scientific and clinical aspects. In selecting Hogness, the IOM touched all bases.[13]

Born in 1922, Hogness was only 48 at the time of his appointment as IOM president, a young man to lead a young organization. He came from a distinguished scientific family. His father, a physical chemist, had taught at the University of Chicago and played a key role in the Manhattan Project that led to the development of the atomic bomb. His brother became a distinguished biochemist at Stanford, and John himself took his undergraduate degree in chemistry. This pedigree no doubt was reassuring to Philip Handler, who could assume that Hogness knew how to function in a community in which scientific achievement was paramount. The physicians at the IOM, for their part, could take comfort in the fact that Hogness had received topflight training at the nation's best medical schools, including the University of Chicago and Columbia, before coming to Seattle in 1950 as a chief resident at King County Hospital, a teaching hospital for the newly formed University of Washington School of Medicine.[14]

"At last we have a pope," one waggish doctor remarked on learning of Hogness's selection.[15] Even as Hogness spoke *ex cathedra* of the IOM and its mission, he realized that the new organization was far from becoming the "established church" in its field. The IOM resembled a movie set with a glossy front and the illusion of depth. Stripped of all the promises about what it would become, the IOM consisted only of a group of members and a president who had the services of one staff member. It would be up to Hogness to make something of an organization that had spent most of its time arguing about its role in the Academy and relatively little time on concrete projects that would, in the end, determine its reputation.

Speaking to anyone who would listen, Hogness tried to interest people in the IOM. He described his new job as "one of the most important jobs in the health field," as head of an organization that "alone in the health field will speak . . . without an axe to grind." Members, who were at the peaks of their careers, would marshal the scientific wisdom of the United States and make recommendations "widely recognized as authoritative."[16] It was important that the organization not play politics, because it could not favor one side over the other and still speak with an impartial sense of authority. In this regard, the IOM would investigate and arbitrate far more than it would assert or advocate. At the same time, the IOM would not be passive and wait for disputes to come to it. Instead, it would seek out problems, looking for significant matters of national policy. "When there are major issues and concerns before the American public, we are more apt to be involved in those areas. . . . We will be able to foresee future problems, rather than deal with them after they have reached crisis proportions," said Hogness,[17] who promised "one hell of a show."[18]

Staffing the Institute and Bureaucratic Routines

Despite this bravado, Hogness told the Executive Committee at the end of 1971 that he had traveled widely in the past year and found few people who were even aware of the Institute's existence.[19] In part, the IOM's anonymity resulted from Hogness's preference to get his own house in order, by hiring the IOM's core staff and establishing its bureaucratic routines, before he reached out to the external community. In September 1971, Hogness announced his intention to hire a senior professional to serve as his deputy. He had in mind Roger J. Bulger, associate director of medical education for allied health at Duke University, with whom he had developed a rapport and who brought distinct skills to the job. If the IOM president came from the West Coast, his executive director would come from the East.[20]

In fact, the careers of Bulger and Hogness were closely intertwined. Bulger, a major figure in IOM history from 1972 to 1976, received his undergraduate degree from Harvard in 1955. After a year in England, he returned to Harvard as a medical student. Bulger took much of his postgraduate training at the University of Washington, serving as chief resident in medicine in the 1964–1965 academic year and, significantly for the IOM, meeting John Hogness. His medical education and postgraduate training in science prepared the way for an academic appointment at the University of Washington. In 1970, he left for Duke, where he held the title of professor of community health sciences and associate professor of medicine. Still only 38 years old, he came to the IOM with a wealth of experience. More importantly, he had a quiet competence and an easy manner that made the work of the IOM go smoothly. He served as the inside man who ran the store while the IOM presidents made outside appearances.[21]

After hiring Bulger, Hogness appointed the members of the IOM's standing committees. In September 1971, he announced the formation of a Program Committee, headed by Irving London, to oversee requests for studies and establish priorities among them. The work of this committee became so important that Hogness decided that it required its own staff. He hired Karl Yordy, the associate administrator for program planning and evaluation at the Health Services and Mental Health Administration, for this purpose. Yordy, who would remain in positions of authority at the IOM from the era of John Hogness until the era of Ken Shine, soon became indispensable not only to the Program Committee but to Hogness and Bulger as well. As for other committee assignments, Dr. Clifford Keene, president of the Kaiser Foundation hospitals, agreed to head a Finance Committee, and Hogness convinced Robert J. Glaser to continue as head of the Membership Committee.[22]

With the basic structure in place, Hogness, turning to the most important committee of all, decided that the Executive Committee was too large to function in this capacity. He renamed it the IOM

Council, and it became the organization's central governing body. From the IOM Council, Hogness drew members who would act as a Report Review Committee, once the IOM began to issue reports. The Executive Committee, meanwhile, was transformed into a much smaller group of only five members.[23]

Hogness realized that one of the most important tasks he and the Council faced was to make plans for the IOM's first annual meeting. The bylaws stipulated that there be such a meeting, and even without a specific mandate, the need for it was obvious. Members required a forum in which they could meet one another, discuss the IOM's program, and become a part of the organization. Hogness set the meeting date for the middle of November and put Samuel Nabrit, Irvine Page, and Eugene Stead in charge of coming up with a program. They devised panels on the founding of the IOM and on "medical care as related to scientific research." In an offhand comment that said a great deal about gender relations at the time of the IOM's founding, the program planners announced that "there will be no special program for wives, and no plans for their attendance at the banquet."[24]

Science reported with some optimism that "the institute seemed to be satisfactorily en route toward establishing an identity of its own" at the November 1971 meeting. The journal reported, however, that the IOM was new and had practically no business to transact. It filled the void with speeches. Walsh McDermott recounted the founding of the IOM. Victor Sidel, head of the Department of Social Medicine at Montefiore Hospital in New York, discussed his recent trip to China. In a concluding speech, John Knowles, general director of Massachusetts General Hospital, offered what was described as an "iconoclastic" analysis of health policy, urging the Institute to get involved in modern problems and not become a "status organiz-ation."[25]

For Hogness, the main objective was to keep the members active and involved. Without much of a scientific product "to offer the membership," he thought it important "to put on a good program and have an outstanding reception." So Hogness and his staff laid on an "very elaborate, excellent buffet." In the years that followed, the IOM fall reception, which attracted a unique crowd of researchers, practitioners, members of Congress and their staffs, and health policy officials, became a key social, professional, and political gathering. With tongue only slightly in cheek, Hogness later wrote that "if, in the first years of our existence, we were not yet too well known, at least we ate well."[26]

Cancer Wars

If the IOM were to succeed, it had to do more than host a pleasant cocktail party. It required a substantive program, something that had largely eluded the Board on Medicine. One of the organization's first opportunities to contribute to medical policy came during the "cancer wars" that dominated the medical news in 1971. President Richard Nixon's decision to declare war on cancer shook up the medical research establishment. Senators Edward Kennedy (D-Mass.) and Jacob Javits (R-N.Y.), neither of whom enjoyed warm relations with the administration, introduced a bill to create a National Cancer Authority, which received wide support from such organizations as the American Cancer Society. The potential issue for the IOM was not whether the federal government should support cancer research but rather how to organize this research. Kennedy and Javits hoped to separate the cancer effort from the rest of the National Institutes of Health in an effort to give it more visibility and free it from the red tape that supposedly hindered NIH efforts. Many IOM members and their colleagues in universities and medical centers felt that such a separation would seriously undermine the NIH and medical research in general. Each special disease or cause would ask to be elevated above the rest of NIH, and soon the entire organization, which had put so much money in medical school coffers, would disintegrate. Nor did creation of a separate National Cancer Authority appear to be conducive to good science. As several IOM members noted in March, "there is the mistaken belief that equates a separate agency with curing cancer."[27]

Philip Handler, whose own background was in biological research that took place in a medical school, felt strongly enough about the issue to write Kennedy and Javits a letter advising against the National Cancer Authority. Handler urged the IOM to move on the issue, but with the organization still so new and eager not to make a mistake in handling its first big political issue, IOM leaders hesitated to make a public stand.[28]

When the political action shifted from the Senate to the House, John Hogness received a request to testify before Congressman Paul Rogers (D-Fla.) and his Subcommittee on Public Health and Welfare. Hogness also was wary of establishing a bad precedent, in this case by testifying as an individual. He preferred to reserve his congressional appearances for times when he would be able to present the results of an IOM report or study.[29] In this case, however, he decided to break his own rules. The cancer agency, he told a reporter, was "a little different because it went to the nature of health research." On

October 4, 1971, Hogness testified before Rogers, and he continued to work behind the scenes with the congressman to create a compromise measure that gave priority to the cancer effort but left the National Cancer Institute within the National Institutes of Health. During the debate on the House floor, Rogers announced that "the presidents of the National Academy of Sciences, as well as the Academy of Medicine of the National Academy of Sciences, have stated strong support of this approach."[30] Rogers got the name wrong, which would be a continuing problem for an organization so closely associated with the National Academy of Sciences, yet whose name sounded so much like the National Institutes of Health. His remarks, however, indicated that the IOM had made an impression on him and on the policy process.

Having played a role in the creation of the cancer program, Hogness looked for a way to influence its subsequent development. The opportunity came in 1972 when the NAS received a request from the Office of Science and Technology in the White House to review the National Cancer Program for the National Cancer Institute (NCI). The idea was for the IOM to evaluate the process by which the NCI developed scientific recommendations and to examine the NCI's management plans. The Institute of Medicine persuaded Dr. Lewis Thomas, a distinguished scientific writer, cancer researcher, dean of the Yale Medical School, and IOM member, to chair the panel that was to write the review.[31]

Working quickly, Thomas and his committee produced a draft report by the end of November. In general, the report praised the National Cancer Program Plan, although it cautioned against leaving the impression that a cancer cure could be programmed in the same way as a mission to the moon. The plan occasionally gave the impression that "all shots can be called from a central headquarters, that all, or nearly all, of the really important ideas are already in hand, and that given the right kind of administration and organization, the hard problems can be solved. It fails to allow for the surprises that must surely lie ahead if we are really going to gain an understanding of cancer." In other words, Thomas and his colleagues wanted to leave the door open for unprogrammed basic research.[32]

The report was unlikely to spark much opposition from the National Academy of Sciences, yet it led to a conflict between the two organizations that reflected the uncertain lines of authority between them. Hogness realized that the report, like all of the reports that the IOM hoped to produce, would have to be reviewed by the Institute. He appointed a committee led by Walter Rosenblith, a professor at the Massachusetts Institute of Technology (MIT) and former Board on

Medicine member who held the distinction of belonging to the NAS, the IOM, and the National Academy of Engineering, to perform this review. Rosenblith's committee posed no objections to releasing the report. The sticking point came over whether the NAS Council would also have to review Thomas's report. Technically, it had such a right, but it often waived that right for something already reviewed by another part of the organization. Hogness hoped that this would be the case with the Thomas report. The decision, Hogness was led to believe, lay with NAS vice president George Kistiakowsky. When Hogness heard that Kistiakowsky thought that formal review could be eliminated, he was relieved, because the IOM faced a tight deadline to get the report to the White House.[33]

Hogness then learned with dismay that Philip Handler had reversed Kistiakowsky's decision. "I still sense a need for review by a Report Review Committee," Handler told Hogness. In a view that reflected his high aspirations for the NAS, Handler wanted to be sure that the IOM group had measured the National Cancer Program Plan against the national interest. He saw the report as establishing an important precedent for the National Academy of Sciences: "The Cancer Plan will surely be followed by the Heart Plan, a Neurological Disease Plan. . . ." Hogness had no alternative but to yield.[34] In the future the report review process would be a major factor in delaying the release of IOM reports, even if it sometimes provided a beneficial check on the reports' quality.

The internal cancer wars at the NAS did not end there. Hogness and Handler then battled over who should transmit the report. Handler thought it was an NAS report and should be transmitted by him. Hogness believed in no uncertain terms that as an IOM product it should be sent to the White House over his signature. "It is essential to the Institute of Medicine that we be recognized as the agency that performed the review and therefore as the responsible agent," he argued. "It is very important that the Institute of Medicine have a distinct identity—that it be regarded as a separate branch of the National Academy of Sciences . . . with its own capabilities and not as the equivalent of another division of the National Research Council."[35]

Hogness won this battle. After he confronted Handler in his office and said that it was simply unacceptable for the report to be anything other than an IOM report, Handler realized that Hogness might resign over the issue. Although Handler yielded, the incident led to a breach between the two men that was never fully mended.

Entitlement and Health Contrasts

Disputes between the National Academy of Sciences and the Institute of Medicine were only one of many reasons that an IOM study might not be completed. Some studies, such as those on collective bargaining in the health care sector or medical responsibilities in criminal processes, languished for lack of funds, despite considerable effort on the Program Committee's part to refine and promote them. Other ideas, such as a request from the National Research Council's (NRC's) Division of Medical Sciences that the IOM study the legal and medical dimensions of "brain death," met with a negative reception by the Program Committee. Still other proposals made it past the Program Committee only to be turned down by the IOM Council.[36]

As the process of selecting studies for the IOM unfolded, it became apparent that the IOM would handle things differently from the Board on Medicine. In the Institute of Medicine, the views of social scientists, who were experts in research design and the testing of hypotheses, carried more weight than they had in the Board on Medicine. The Board on Medicine relied on outside consultants to conduct its research because it was small and compact and had almost no staff at its disposal. Because the IOM aspired to be larger and more comprehensive, it had the potential of using internal staff, rather than external consultants, to undertake a broad range of studies. It also hoped to be agile and quick enough to do studies that were relevant to current policy concerns. Achieving these various goals proved very elusive, as the rise and fall of the entitlement study demonstrated.

The Institute of Medicine inherited the entitlement study from the Board on Medicine. Board members had recognized national health insurance as one of the most important concerns of health care policy and wanted to do something about it. Negotiations with federal authorities for a contract to study the subject stretched into the period after the founding of the IOM. Only then did Paul Sanazaro, director of the National Center for Health Services Research and Development—a federal agency in the Department of Health, Education, and Welfare that was started at about the same time as the Board on Medicine and for many of the same reasons—ask the IOM to undertake "a study in depth of universal entitlement" to health care. Such a study would consider "fundamental questions and issues related to the structure and implementation of a national program of entitlement."[37]

This mandate was breathtakingly broad. It amounted to nothing less than planning a program of national health insurance and explaining what effects it would have on the nation's health. It reflected the sense in the early 1970s that passage of national health insurance was inevitable, and therefore the more rational, dispassionate planning that preceded it, the better. The study was part and parcel of the Board on Medicine's outlook that it should engage broad social questions.

Although Sanazaro's request was made in January 1971, it took until June for the IOM to obtain a small planning grant, designed to produce a larger, more elaborate proposal, from the Department of Health, Education, and Welfare (HEW).[38] There was considerable interplay between Philip Handler and the leaders of the IOM in the process. The trouble was that much of the work for the project was to be done by Leon White, a professor of operations research and management at MIT. In March 1971, when Bryan Williams, the Dallas physician and former Board on Medicine member, went before the NAS Council to explain the project on behalf of the IOM, he met with a barrage of criticism so fierce that Handler felt compelled to apologize to Robert J. Glaser. Handler told Glaser that "it will not do to have the study conducted as an all MIT affair; somehow a multidisciplinary committee of the Institute must be involved in the planning, participate in some part in the conduct of the study, and take absolute responsibility for the final report." Glaser said that he understood the problem and reassured Handler that White would be carefully supervised and controlled, both by Bryan Williams in Dallas and by Rashi Fein, Walter Rosenblith, and Irv London, who were on the scene in Boston.[39]

White, it became clear, wanted the study to be done in a manner similar to the Carnegie Commission Study on the Future of Higher Education. He expected a commission to be formed by the IOM that would authorize the conduct of specific research projects. Hogness told White that the NAS was very "leery about subcontracting when the subcontract implies a policy making activity or judgment activity on the part of the subcontractor." "I share this reservation completely," Hogness added.[40]

The planning group for the entitlement study, headed by Bryan Williams, met for the first time in August. By September, the group had come up with a proposal for a full-fledged study that would cost $2 million. The IOM Executive Committee approved the proposal by a mail ballot, although not everyone appeared to be in favor of pursuing the study. Dwight Wilbur, former president of the American Medical Association and a distinguished private practitioner, thought that the

18 months allotted to such a large study was insufficient. He hoped that the private sector would be emphasized as much as the public sector. Rashi Fein, an academic economist and avowed partisan of national health insurance, was concerned that planning for national health insurance would be suspended for the life of the study, thereby delaying passage of the measure.[41] These comments revealed that the project had begun to encounter the polarized politics of national health insurance.[42] Nonetheless, the NAS Council joined the IOM Executive Committee in approving it.

Just as this proposal was ready for public distribution, the IOM's Program Committee began to meet. When it took a fresh look at the proposal, more doubts began to surface. For one thing, members felt they should have been consulted. If they had been, there might have been more in the proposal about the place of health technology and the role of health manpower. For another, there was a sense that the study was too broad.[43] Among the most vocal critics of the study was David Mechanic, a leading figure in the sociology of medicine who had entered the IOM in the summer of 1971. He said that the project would inevitably result in disappointment because it promised more than it could reasonably deliver. The proposal attacked "almost every issue in the health services research field, many of which are unanswerable at this time."[44] At base, Mechanic criticized the IOM for displaying a naivete about social science research.

In February 1972, a group headed by Irving Lewis, a professor of community health at the Albert Einstein College of Medicine, met in Boston to see if the proposal could be saved. William Schwartz, chairman of Tuft's Department of Medicine, stressed again that the study was attempting to address too many questions. He questioned whether IOM members would have the time to supervise such an ambitious study. Julius Richmond, the head of the Judge Baker Guidance Center in Boston who was destined to play a major role in IOM history, asked whether the study would force the Institute to take a position on national health insurance. Roger Bulger replied that Hogness hoped, at the very least, that some general principles for national health insurance would emerge. This posed a further dilemma. If the study was broad, it would require a long time to complete. If the study was to be relevant to the political debate, it would have to be finished quickly. After the meeting, Lewis conceded to Hogness that "I just do not see how we can proceed with the present scope of work."[45]

Because the members did not seem to want the study, Hogness decided to kill it. Martin Feldstein, the prominent Harvard economist and an IOM member, concurred in this decision, advising Hogness that if the IOM had done the study, it would have risked becoming a "political body hiding behind some not very good analytical work." Feldstein's low opinion of the proposal reflected his lack of confidence in the analytic skills of the Institute staff and the Program Committee. Neither, Feldstein believed, could "judge the quality of prospective social science research." According to Feldstein, the IOM proposal would not have been approved by the National Science Foundation, which funded work in economics and other social sciences.[46] If Hogness's decision to kill the project marked a response to internal IOM intellectual politics, it also marked a response to the external political environment. Hogness knew it would be difficult to obtain funding for a broad study of health insurance because the administration had submitted a health insurance proposal of its own. The National Center for Health Services Research and Development, a creature of the Great Society, had long since fallen out of favor with the Department of Health, Education, and Welfare. Hogness learned from a high-ranking HEW official that the department had no desire to sponsor a study that might contradict what it already had done. The window for the IOM to influence the debate was closed.[47]

The study on universal entitlement taught Hogness two important lessons. First, IOM studies would have to be closely supervised and not entrusted to outside contractors. "As you know," Hogness wrote Feldstein, "we inherited the Universal Entitlement Project from the Board on Medicine, and they had a somewhat different attitude about the role the Institute members should play in supervising such studies." Second, the organization would have to develop the technical capability to critique proposals, particularly in the social sciences areas.[48] As a first step toward implementing the lessons, Hogness decided to rid the Institute of the inventory of old Board on Medicine projects and to proceed with new ones. These would of necessity be smaller in scope than the old projects and would rely more on the Institute's medical, rather than social science, expertise. If the IOM should undertake a large study, it would do so with its own staff, rather than subcontracting the bulk of the work to outsiders.

The chief legacy from the Board on Medicine was a series of studies conducted by David Kessner, an internist from Yale University. These studies examined contrasts in health status between the rich and the poor, showing how differences in health outcomes were related to differences in health care delivery. Financed by the Carnegie Corporation and other foundations interested in problems of poverty, the project involved, among other things, field studies in which teams of doctors examined children living in low-income Washington neighborhoods. Walsh McDermott saw this project in typically grandiose terms. It would "appraise the extent to which the health problems of the disadvantaged are a consequence of economics, race, failure of the delivery system and a failure of the other related systems in society." Although the results were much more limited, the study did yield useful methodology for using certain conditions, such as anemia and middle-ear infections, as statistical "tracers" that could be employed to compare the effectiveness of different health delivery systems. The project also mined data sets such as a record of live births and infant deaths in New York City in 1968 to show that differences in infant deaths between whites and blacks could be somewhat reduced through "easily obtained personal and medical information."[49]

The health contrast studies resembled the sort of work that a faculty member might do in a school of public health. In sponsoring them, the Institute of Medicine put its imprimatur on a research project that was largely conducted by outsiders. The Institute did not use the studies to make authoritative statements on health policy questions, in part because IOM Council members such as Harvard statistician Fred Mosteller urged the IOM to be cautious about pushing its conclusions beyond the reach of the data.[50]

Health Effects of Abortion

In contrast to the entitlement and health contrast studies, the abortion study showed how the IOM could contribute to health policy debates, even on a controversial topic. This project stemmed from the famous January 1973 Supreme Court decision in the case of *Roe* v. *Wade* that states could not ban abortions during the first trimester of pregnancy. In the fall, David Hamburg and David Mechanic, both of whom were IOM Council members, suggested that the IOM might want to examine the health impact of legalizing abortion. The idea captured the imagination of the IOM Council and won the approval of the Program Committee, the Council, and NAS authorities.

In February 1974, John Hogness invited Dr. Mildred M. Bateman, director of the West Virginia Department of Mental Health, to chair the steering committee created for the abortion study. Ten others joined her, including seven IOM members. All came with imposing titles: the head of the Obstetrics and Gynecology Department at Meharry Medical College, the chief of the Epidemiology Department at Harvard's School of Public Health, the dean of the Medical School at Case Western Reserve University, the president of the Social Sciences Research Council.[51] The committee's pedigree testified to the

IOM's unique ability to assemble an interdisciplinary panel of distinguished medical and social sciences practitioners.

To run the study, the IOM needed to raise money from outside funders. It turned to a group of private foundations, each of whom made small grants of about $10,000. For the seven groups that said yes, at least as many said no. This made fund-raising a time-consuming, frustrating activity that continued throughout the project. When the project began, no one knew if the money would hold up long enough for it to be completed.[52]

Unlike the entitlement and health contrast studies, the steering committee for the abortion study made no effort to perform original research. Instead, it saw its mission as synthesizing and critiquing the existing literature. Every two months, committee members flew into Washington, D.C., for meetings. Nearly all of them, burdened with busy schedules, missed a meeting or two. Inevitably then, the steering committee looked to Martha Blaxall, an IOM staff member who held a doctorate in economics, to draft the report. She functioned as an adjunct of the steering committee, not as an independent contractor, as David Kessner had on the health contrast studies. In the parlance of medical research, she was not the principal investigator who ran the study as much as the talented postdoc working in someone else's laboratory. Just as science was a collaborative endeavor, so the line between the steering committee and the staff on an IOM project was difficult to draw. Although the steering committee reviewed Blaxall's work carefully and came to its own conclusions, the study nonetheless depended very much on her efforts.[53]

Because the steering committee met infrequently and internal clearances were necessary for any IOM report, the report on legalized abortion was not released until May 1975. At the very beginning, the committee made it clear that the report would not deal with the ethics of abortion. It would, instead, treat the health aspects of the subject, an area in which its credentials were unquestioned. For many combatants in the debate, this posture resembled mounting a production of *Hamlet* without the character of Hamlet. From the IOM's perspective, the decision showed a newfound ability to limit its work to those aspects of public policy on which it could speak with authority. For the members of the steering committee the ethics of abortion were, in any case, beside the point. They believed that whether or not abortions were legal, women would seek them. The committee concluded that "legislation and practices that permit women to obtain abortions in proper medical surroundings will lead to fewer deaths and a lower rate of medical complications than restrictive legislation and practices."[54]

In a manner typical of IOM study groups, the steering committee conceded that many things about abortion remained unknown. Hence, in what would become a cliché in IOM reports, the group called for further research on the consequences of abortion on health status. The highest priority, according to the committee, was for studies of the effects of abortion on mental health.

Although this IOM lacked the money for wide distribution of the report, it nonetheless elicited considerable public reaction. The Center for the Study of Moral Order condemned the committee for not taking a stand on the moral and ethical issues and urged the IOM to "disavow the ethically neutral behavioral movement of social numbers theory."[55] Wire services wrote stories about the report that led to newspaper articles from coast to coast. The report's summary and conclusions made their way into the *Congressional Record* on June 3, 1975. The National Abortion Action League and the Religious Coalition for Abortion Rights distributed reprints from the *Congressional Record*.[56] In this manner, although the study contained nothing surprising and tended to reinforce the conventional liberal wisdom, it made a definite impression on the public debate. The study showed that the IOM could work in a modest manner and still get results.

Policy Statements

The IOM also experimented with small-scale policy statements on timely issues. These were short reports done quickly. One of the first came about as a result of a provision in the Social Security Amendments of 1972 that awarded health insurance coverage through Medicare to people with end-stage renal disease (ESRD). The idea was to have the federal government pay for kidney dialysis. The IOM quickly convened a group to consider whether this disease-by-disease approach to national health insurance made sense.

In a sense, the composition of the group predetermined its outcome. At least three of the members followed Social Security policy closely and objected to the way in which Congress had handled the ESRD provisions. Robert M. Ball, a scholar in residence at the IOM who served as Social Security Commissioner from 1962 to 1973, acted as a consultant to the committee. He, too, opposed the ESRD parts of the 1972 law. Not surprisingly then, the panel, chaired by political scientist Herbert Somers of Princeton University, concluded that "coverage of discrete categories of diseases would be an inappropriate course to follow in the foreseeable future for providing expensive care on a universal eligibility basis."[57] In other words, the study reaffirmed the convictions that panel members brought to the task—a potential danger inherent in all IOM reports and, for that matter, in all NAS reports.

The IOM intended the health maintenance organization (HMO) study to function in a manner similar to the categorical approach report. It would be a modest effort that would produce what John Hogness called a "background paper on health maintenance organizations." Hogness discovered, however, just how hard it was for the IOM to stay current with breaking events.

The committee for this project, headed by Paul Ward, executive director of the California Committee on Regional Medical Programs, met for the first time on January 9, 1973. Everyone realized that federal legislation in aid of HMOs was under consideration in Congress. Despite the possibility that the committee's work might be overtaken by events, the group decided to proceed. The committee produced a preliminary draft by October; President Nixon signed the HMO Act of 1973 into law at the end of December. The law implemented some of the committee's suggestions, yet the committee continued to see the merits of issuing its own report.[58]

The theme of the report was that health maintenance organizations should be given what the committee called a "fair market test." Such a test would "do more to improve the functioning of the American health care delivery system than any other policy step which could be taken in the near future."[59] IOM members who reviewed the report were less than impressed with it. One reviewer called the timing of the report ". . . peculiarly unfortunate. It is both too late and too soon." The IOM could no longer influence the content of the law, nor could it appraise the law's impact. Furthermore, the committee's fair market test, according to the reviewer, came complete with numerous regulatory suggestions that amounted to a "high degree of protection." Such criticism led to further delays in releasing the report. One federal official had to plead with the IOM's Karl Yordy to get the report released. "Let's get the Report published and let's forego any further nitpicking," he said. When the report appeared in May 1974, it had only a minimal impact on public policy.[60]

The Costs of Education Study

The reports on HMOs, end-stage renal disease, and abortion required a large expenditure of staff time with comparatively little return. The reports were on disparate topics and failed to bring a sense of coherence to the IOM's efforts. If the IOM were to survive, it required a large sustaining project that brought in enough money to cover not only the costs of the project itself but some of the Institute's other operational costs as well. The study of the costs of educating health professionals served as just such a project for much of 1972 and 1973.

In November 1971, John Hogness received a letter from Merlin K. DuVal, Assistant Secretary for Health and Scientific Affairs at the Department of Health, Education, and Welfare, mentioning the fact that Congress was about to pass the Comprehensive Health Manpower Training Act of 1971. DuVal had managed to insert a provision in the act that mandated a study of the "national average annual cost of educating students in each of the health professions." DuVal made sure that the Institute of Medicine would be asked to perform the study.[61]

The study concerned a subject in which IOM members took a vital interest. At stake was the way in which the federal government would subsidize medical schools as well as schools of dentistry, osteopathy, nursing, veterinary medicine, pharmacy, podiatry, and optometry. In the past, much of the support was indirect, through mechanisms such as research grants, which by 1968 amounted to nearly 42 percent of medical school revenues; construction grants for teaching faculties; and special grants for schools in financial distress. With federal research funds declining and medical school enrollment expanding, Congress sought a better system for aiding schools. One idea was to make grants to the schools based on the number of students. Not ready to make a final decision on how to implement such a system, Congress fell back on the old device of requesting new data. It wanted to know how much it cost to educate a medical student and how much variation existed from school to school. It also asked for a way in which costs per student could be calculated and for any recommendations that the IOM might have on the way the federal government could use educational cost data to determine the appropriate amount of capitation grants to health professional schools.[62]

The assignment required the IOM to come up with hard data in a manner similar to the studies on health contrasts that had been conducted by David Kessner. This necessitated hiring a large staff with technical competence and the ability to complete the study by the June 1974 deadline imposed by Congress. Hogness put Ruth Hanft, the IOM's senior research associate, whom he had hired in March 1972, in charge. He had met her during his service on the HEW Secretary's Task Force on Medicaid and had come to admire both her technical skills and her Washington savvy. Hanft was, in fact, a distinguished figure in health services research who, among other things, enjoyed a close working relationship with HEW Secretary Elliott Richardson and had played a key role in designing the Nixon administration's health insurance proposal. Hanft, in turn, supervised the hiring of a staff that eventually numbered 20 people, as well as consultants from five different consulting firms.[63]

The costs of education project was a $2 million undertaking that involved reviewing data from previous studies, coming up with a methodology for estimating costs, gathering the data necessary to make the estimates, interpreting the data, and producing a final report. Hanft and her study team decided to appoint eight professional advisory panels, one for each of the professions being studied. This necessitated holding meetings with each of them. The most time-consuming aspect of the project involved visiting a representative sample of the schools and then asking faculty members to record their activities for a week of their time. In this way, the project staff hoped to determine exactly how much of a faculty member's time went into teaching and other educational activities and how much went into providing care or performing research. The approach raised complex methodological issues. How could one reasonably divide the time of a resident who both supervised medical students and interns and performed clinical procedures on patients? How could one be sure that faculty members reported their time accurately or participated in the survey at all? Where, exactly, did teaching stop and research begin, when the research was often conducted with students and when the products of the research informed faculty members' teaching?[64]

Julius Richmond headed the IOM steering committee that tried to advise Hanft and her staff on these issues and formulated the final recommendations. The committee also included Eli Ginzberg, a professor of economics at the School of Business, Columbia University, and one of the nation's leading experts on manpower issues. Almost alone of the members of the steering committee, he devoted a substantial amount of his time to the committee's work and closely critiqued the staff's methodological approach. Others on the committee had direct experience with administering a school concerned with the health professions—for example, James Kelly, former vice president for administrative affairs at Georgetown University; Alvin Morris, vice president for administration at the University of Kentucky; Martin Cherkasky, director of Montefiore Hospital; and David Rogers, former dean of the Johns Hopkins Medical School and current president of the Robert Wood Johnson Foundation.

Inevitably, the steering committee ceded a lot of ground to Hanft and her staff, both because committee members could not hope to get involved in the finer points of data collection and because Hanft was an accomplished expert in her own right. Indeed, she would become an IOM member after her service on the IOM staff. When it came to making recommendations, however, the steering committee played a central role.

The climactic meeting of the steering committee took place in September 1973. The central issue concerned whether to restrict the report to hard data, such as a statement that the total cost of educating a medical student was $4,821, or to broaden the report to include a philosophical statement on the government's role in health education. Daniel Tosteson, the chair of the Department of Physiology and Pharmacology at Duke, hoped to stick close to the data. David Rogers disagreed. He argued that the "climate of the times" demanded the report go beyond costs. The figures were important to support the committee's position, but "if the Institute of Medicine doesn't take a lofty view of Federal subsidy to health professionals, no one will." He concluded that "Congress wanted more than data; they want strong, constructive guidelines." Eli Ginzberg agreed that the report should convey the fact that federal subsidies for educating health professionals were "in the national interest."[65]

The final report, like nearly all IOM reports, amounted to a compromise between these two positions. It included a great deal of data and even a technical appendix. It also contained a series of broad philosophical recommendations, for example, that "health professional schools be regarded as a national resource requiring Federal support." The group endorsed a capitation grant program as "an appropriate Federal undertaking to provide a stable source of financial support for health professional schools" and recommended that capitation grants ranging between 25 and 40 percent of net educational expenditures "would contribute to the financial stability of the schools and would be an appropriate complement to other sources of income."[66]

When it came time to brief the Subcommittee on Health and the Environment of the House Interstate and Foreign Commerce Committee, Hanft, rather than Julius Richmond, did the talking. She dwelled on the methodological aspects of the study and presented the congressmen and their staff with tables of data. She mentioned the group's recommendations last, in contrast to the formal report, which listed them first.[67]

The report on the costs of medical education illustrates some of the problems that the IOM faced in its early years. The study, although large and lucrative, lasted only a short time and originated in Congress, not in the IOM itself. It was not a project that IOM members would have chosen to do on their own. The National Academy of Sciences had a long tradition of responding to requests for advice from the federal government and expected the IOM to do the same. The IOM, for its part, hoped to maintain an independent agenda that reflected professional and societal concerns. The costs of education study furthered this agenda but only indirectly. It concerned a limited and highly technical point of policy, not a major policy initiative. Furthermore, the restricted duration of the study meant that it provided only temporary financial relief to the IOM. The large staff recruited for the project would not be able to move on to other IOM projects. Finally, the project was staff-driven. Steering committee members did little to influence the study methodology, which could as easily have been done by a Washington think tank or consulting firm as the IOM.

What the IOM added to the costs of education study and to health policy more generally was the ability to convene groups of experts to serve on the steering and advisory committees. The appeal to Congress was that the IOM represented a broader interest than the medical and other health professions schools. In this regard, the IOM differed from the Association of American Medical Colleges or other groups that had done studies on the costs of medical education.

In the end, however, the Institute of Medicine contained a core constituency of faculty members and administrators from academic medical centers. On the 10-person steering committee, only Eli Ginzberg, a Columbia professor, and Morton Miller, executive vice president and chief actuary of the Equitable Life Assurance Society, were not directly involved with medical education. The other eight, with the possible exception of foundation executive David Rogers, depended on federal subsidies to medical, nursing, or dental schools for their livelihoods. Hence, the general principle of federal aid for medical education never came under heavy attack. Although Hanft might have preferred to stick with the data and not take any philosophical positions, she was overruled. In the ensuing discussion, there was no conservative ideologue who could hold his or her own with former medical school dean Rogers or federal manpower program advocate Ginzberg. Nevertheless, the IOM delivered an innovative and technically competent report on time. When Senator Edward Kennedy offered amendments to the Health Professions Educational Assistance Act of 1974, he described the IOM study as the "best and most reliable data . . . on the educational costs of schools of the health professions."[68]

The Program Committee

Beyond the Washington Beltway and beyond the large medical centers, few people worried about reliable data on an arcane subject. As a consequence, the costs of education report did little to increase the IOM's general visibility, nor did it help unify the IOM's program. The study, in short, was not the sort of project that Hogness favored as a model for the IOM. He preferred that the Institute not conduct research but rather that it be concerned with "broad and basic issues of national health policy."[69] His advisers concurred in the recommendation that the "Institute concentrate more on policy analysis than on the gathering of original data."[70] In this vein, Hogness and the IOM Council reacted with skepticism to a proposal that the IOM function something like the Brookings Institution. If the IOM were to be a think tank, it would be a staff-driven, not a member-driven, organization, which appeared to violate one of the IOM's main organizing principles.[71] Neither Hogness nor the IOM Council wanted to leave the IOM to the mercy of external clients, even those of such obvious influence and importance as the U.S. Congress, and make it, in the Washington vernacular, a "job shop." Instead, Hogness recognized the need for the organization to focus its efforts on important issues of its own choosing. He looked to the Program Committee to provide the necessary focus. In this manner, he followed the plan set down by the Institute's founders who mandated that the IOM president create a Program Committee and use it to prepare a program of studies.

The Program Committee under the direction of Irving London and with the staff support of Karl Yordy held frequent meetings not only to consider specific proposals but also to produce a general framework for the IOM's work. As a first cut, the committee divided the proposals into 10 general categories. These categories failed to satisfy everyone. James Shannon, still following the IOM's activities with interest, said that the categories gave short shrift to science and medical education. He saw in the committee's work "the reflection of a progressive detachment of the drive of medical schools from a biomedical science base to softer social considerations." Shannon, in other words, was still conducting the war that he had waged in the Board on Medicine against Walsh McDermott. For Shannon, there was too much on "legal and ethical issues" (category 10), "health maintenance" (category 6), and "environmental influences on health" (category 2). Philip Handler found himself in full agreement. Science always came first with him. Roger Bulger conceded that the categories were rather loose but thought it important just "to get started." Even the Program Committee did not regard the categories as fixed. After consulting with members, for example, the committee realized that "government regulations and health administration" (category 9) intersected with nearly all the other categories and would have to be dropped.[72]

At the end of 1972, the Program Committee took the process a step further and issued a general statement that suggested four basic principles to guide the IOM program. These followed directly from the deliberations that had led to the IOM's founding. The organization should rise above "any particular interest, viewpoint, or profession" in favor of "the protection and advancement of the health of the public." The Institute's program should involve its members in its work. The IOM should recognize a "fundamental unity of health policy issues" and not attempt to attack them piecemeal. Finally, the "Institute should initiate activities as well as respond to external requests." There followed a long laundry list of projects that the IOM might undertake, such as an "analysis and redefinition of health manpower functions and roles" or a "study of the interaction between the physicians and the FDA on drug utilization."[73]

Fund-Raising

A fundamental lack of money kept the IOM from following its general principles and undertaking these specific projects. Fund-raising, so vital to the IOM's survival, became a source of considerable anxiety during the IOM's first 15 years. "The financial support of the Institute of Medicine was, to put it mildly, a bit shaky at first," Hogness recalled.[74] In time, however, Hogness attracted a core group of five foundations that came to the IOM's aid. For the first few years, the IOM included a statement in each of its studies that it received its principal funding from the five foundations and listed them by name.

During the long period of discussion with the National Academy of Sciences, the Board on Medicine had put its fund-raising efforts on hold, with the result that the IOM had little to spend in its first months of operation. Through the end of the 1972 fiscal year, the National Academy of Sciences covered most of the IOM's $2,000 in administrative expenses. Only the studies on health contrasts and the costs of medical education brought in much external money, and most of this went to pay specific project expenses, leaving nearly nothing for activities that might lead to the IOM's expansion.[75] As a consequence, the IOM maintained a versatile yet small and rudimentary staff.

Hogness realized that much more money was required. He estimated that the IOM would need at least $700,000 just to sustain its central activities including staff salaries and money to pay the travel expenses of the members of the IOM Council and the IOM Program Committee.[76] Although those who served on IOM committees did so without compensation, they still generated bills for transportation and lodging. Looking to the foundation community to support the IOM, Hogness made the rounds of foundations that took an interest in medicine and medical policy.

One of his first stops was the Robert Wood Johnson (RWJ) Foundation. At the end of 1971, Hogness told the head of this foundation, which was just getting organized, that he planned to make a formal proposal at the beginning of the next year. At the time David Rogers, who was himself an IOM member and regarded Walsh McDermott as an important mentor, would become the president of the foundation. Hogness and Rogers were on a close, first-name basis, with Rogers even relying on Hogness to make suggestions for appropriate people to staff the foundation. Although foundation officials asked many hard questions about the IOM's focus and fund-raising plans, the RWJ Foundation decided to make a major grant of $750,000, to be spent over a three-year period. The foundation predicted that the IOM would "make a contribution of the first importance to the outcome of the difficult and decisive policy decisions confronting the nation's health enterprise." The grant, the sixth largest that the foundation awarded in 1972, represented a substantial vote of confidence in the IOM. The foundation and the IOM hoped, in effect, to work together in establishing themselves as important entities in the world of health policy.[77]

Support from the Robert Wood Johnson Foundation had an immediate effect on IOM activities. In fiscal year 1973, RWJ money amounted to more than the amount contributed by NAS to the organization. The IOM used the money to defray some of the expenses that it could not charge to project accounts, such as the $157,000 it spent on salaries for "IOM direction and support," the $40,000 required to fly the Council to Washington, D.C., for meetings, and the

$46,000 that it hoped to spend on public information and reports.[78] All in all, the IOM used $200,000 of the foundation's money in 1973. Hogness credited the RWJ Foundation with making it possible for the IOM to expand from 5 to 12 professional staff members.[79]

The W.K. Kellogg Foundation, which had a long-standing interest in public health, provided another target for Hogness's philanthropic outreach. Here again, it took substantial effort for the IOM to satisfy the foundation. A proposal calling for $643,420 over five years was rejected by the Kellogg Foundation and led the IOM to write a second proposal, asking Kellogg to give the IOM $100,000 a year for five years. IOM officials estimated that this amount would cover 40 percent of the basic operating expenses. Only a month after the RWJ award, the Kellogg Foundation board elected to give the IOM $100,000 a year for three years, a grant that was not as long in duration as the IOM had requested but one that paid just as much each year. The foundation hoped the IOM would use this money to plan its major studies and to develop "authoritative statements on major health policy issues." Hogness reported that this "flexible support" would enable the Institute "to take the initiative" on such issues as the "training and distribution of health manpower." The IOM used the money to fund short-lived member "survey" committees to assist the Program Committee, such as one on science policy for medicine and health, and to pay for staff papers on such subjects as the Nixon administration's 1974 health budget.[80]

The Richard King Mellon Foundation became the third foundation to come to the IOM's aid in 1972. Robert J. Glaser, who worked at the Commonwealth Fund, reminded John Hogness that during the Board on Medicine days, Irvine Page had gone to see a member of the Mellon family and asked for "a massive amount of money to set up an independent academy." Glaser urged Hogness to visit George Taber, director of the Richard King Mellon Trust, and tell him that the IOM was the direct outgrowth of Page's efforts. Hogness followed through on Glaser's suggestion, and this yielded a grant of $300,000, to be spent over three years, in support of the IOM's budget for central activities. Most of the money went to pay staff salaries.[81]

By 1975 two more foundations, the Commonwealth Fund and the Andrew Mellon Foundation, had joined the core group of major contributors to the Institute of Medicine. Getting money from both required intensive efforts on the part of Hogness and his staff. Commonwealth turned down an initial request for $1 million. Hogness persisted, submitting another application at the end of November 1973. The Commonwealth board, which included IOM members Quigg Newton and Robert Glaser, deliberated over this proposal until May. During this period, IOM officials made numerous visits to Commonwealth and talked with its officials by telephone. The effort paid off in the form of a grant of $200,000 a year for three years.

Cultivating the Andrew Mellon Foundation proved similarly arduous. There, Hogness dealt with Nathan Pusey, the former president of Harvard, and John Sawyer, the former president of Williams. He asked Mellon for a "spendable endowment grant of $1,000,000." Mellon officials were not interested in this. Their foundation tended to focus on the humanities, rather than the sciences. In time, the IOM developed a proposal to establish a program in health care ethics, something that fit the foundation's mission and had been on the IOM's agenda from the beginning. In December 1974, Pusey informed Philip Handler that the foundation would give the IOM $750,000 for its program of "examining the competing political, economic, and social values inherent in decisions pertaining to health care policy."[82]

Although preferable to receiving money from the federal government to perform a specific task, foundation funding carried its own drawbacks. For one thing, foundation officials—some of whom were IOM members and hence entitled to kibitz for that reason alone—enjoyed a relationship with the IOM not unlike that between New York financiers and Hollywood producers during the studio era. In return for their money, the foundations expected to exercise a certain degree of oversight or, in the language of philanthropy, stewardship. It therefore mattered when a high official of the Commonwealth Fund criticized the IOM for taking a "shotgun" approach to health policy issues or an official from the Kellogg Foundation said that the IOM did not do enough to disseminate the results of its studies.[83] There was, in other words, no free lunch.

It also took considerable industry to "scrounge up" the lunch. Most of the philanthropic contacts that Hogness made went nowhere. With such an illustrious group of members, the IOM could get in touch with nearly any foundation and meet with a cordial reception. Often, however, the IOM received exquisite courtesy and nothing else. The case of the Ford Foundation was typical. Adam Yarmolinsky knew McGeorge Bundy, head of the Ford Foundation, through both his Washington and his Harvard connections, and he set up a meeting with Bundy; Ford vice president "Doc" Howe, a former commissioner of education; and IOM officials. "My guess is that ours is not the particular bank at which they will be most likely to find anything but good wishes," Bundy told Yarmolinsky. At the meeting, Bundy offered only a glimmer of hope, advising the IOM that the Ford Foundation was not in medicine, gave no money for endowments, and was not feeling rich. For the next eight months, IOM staff members kept plying Bundy with invitations to IOM events and updates on IOM activities. Then the Ford Foundation politely but firmly said no.[84]

Perhaps the most difficult thing about foundation grants was that they had to be renewed. When IOM officials received a foundation grant, they spent, rather than invested, the money. This meant that within a few short years, they had to return to the foundation or find another one to take its place, a situation that did not make for financial stability. In the common parlance, foundations provided soft, not hard, money. "I am still concerned about the long-range financing of the Institute. Foundations tend to be fickle in their support," wrote William Danforth, a physician, IOM member, and president of Washington University, in 1975.[85]

Organizational Routines

Even with foundation support, the Institute of Medicine could do only a few of the things it wanted to do. To create a sense of community, the Institute looked to activities that were under its control. Of all the organization's rituals, members took the most interest in the annual election of new members, although the results never pleased all of the IOM's constituencies. The elections between 1972 and 1975 were typical in the way that so many members participated and so many were discontented with the results. In 1972, the members made 150 formal nominations for 50 available positions, and the Membership Committee reduced the number to 84. The election yielded 53 new members, including former HEW Secretary Wilbur Cohen, philosopher and medical ethicist Daniel Callahan, distinguished sociologist Elliot Freidson, and noted health statistician Dorothy Rice. With the results in, the IOM Council analyzed them to identify areas that had to be strengthened. Its analysis showed that only 4 of the 17 practicing physicians nominated, compared with 10 of 11 academics, made it into the Institute, and that only 4 basic scientists were elected. Less than half of the 25 new members had an M.D.; 12 were members of minority groups, including 10 blacks. The results stirred up Irvine Page, who complained that the IOM was becoming a "sociological forum far from the needs of medicine. . . . We are increasingly being looked upon as academicians of strong political and social hue."[86]

Hogness agreed that more practicing physicians and more basic scientists were needed. Although the Institute made a concerted effort to elect more basic scientists in 1973, fewer scientists were picked than any other group. This agitated Philip Handler, who suggested that there be "fixed quotas of relatively non-overlapping categories" in order to balance the membership. The Membership Committee came up with a new set of categories, including clinical and fundamental scientists, that resulted in the election of a few more scientists. This, in turn, irritated social scientists such as David Mechanic, who complained that only one of the 33 candidates listed as a fundamental scientist—economist and NAS member Kenneth Arrow—was a social scientist. No psychologists or anthropologists made it on the list. In succeeding years, the Membership Committee continued to adjust the categories and created a ballot that set quotas on each category. In 1975, for example, the categories included natural scientists, social scientists, health administrators, clinical practitioners, and others. Members were urged, although not compelled, to vote for a specified number of candidates in each category, up to a total of 40.[87] The result was that the proportion of fundamental scientists rose from 14 to 15 percent of the membership, but practitioners declined from 9 to 8 percent.[88] The search for the proper mix of members continued.

If elections provided a common bond among Institute members, the presence of visiting fellows helped to create a sense of community in IOM central offices in Washington, D.C. Most came with outside money and sought only a congenial place in which to work. At the beginning of April 1973, former Social Security Commissioner Robert Ball moved in as a more or less permanent fellow. He remained at the Academy until the beginning of the Reagan era, using his time to write a book on Social Security and Medicare and to play an active role among Social Security and Medicare policymakers. With his tremendous set of Washington connections and extensive administrative experience, he served as an informal adviser to at least four IOM presidents. As an IOM member in his own right, he also took a more formal part in governance, serving as a member of the Council and of the steering committees for IOM studies.

Dr. Robert Q. Marston, former director of NIH, became a fellow at the same time as Robert Ball. Although his stay was brief, his mission was important.[89] He directed the Robert Wood Johnson Health Policy Fellowships program, which became the longest-running regular activity of the IOM. It began with a $710,000 grant from the foundation to the IOM, intended to "offset severe shortages of faculty members in the nation's academic medical centers who are specifically qualified for research, teaching, and service in the complex field of health policy."[90] The idea was to bring a small number of academic physicians and other health professionals to

Washington and have them work for congressional committees and other agencies involved in making health policy. Cooperating with the American Political Science Association, the IOM ran the program through a special Fellowships Board that took the lead in selecting the fellows. The Institute announced the selection of the first group of six fellows, who had been culled from 43 nominations and 12 finalists, in the spring of 1974. These fellows spent the 1974–1975 academic year in Washington, starting with an orientation at the IOM and then moving into temporary assignments on Capitol Hill. David Banta, an assistant professor at Mount Sinai School of Medicine's Department of Community Medicine, with both an M.D. and a master's degree in public health to his credit, worked for Paul Rogers's health subcommittee in the House and for Edward Kennedy's health subcommittee in the Senate. He thought enough of the experience to accept a permanent Washington assignment.[91]

Although foundation officials worried that having the fellows succumb to Potomac fever defeated the purpose of the enterprise, they recognized the fellowship program as an important endeavor that deserved their continuing support. In 1976, the RWJ Foundation renewed the basic grant and raised its level to more than $1 million. The fellows, meanwhile, had become a regular part of IOM life. They received standing invitations to the IOM annual meetings and other activities.[92] Despite occasional squabbles over budgetary levels and administrative details, the fellowship program exemplified an almost seamless collaboration between the RWJ Foundation and the IOM, with few of the drawbacks of other forms of soft money.

Relations with the National Academy of Sciences

If annual meetings, elections, and visiting fellows provided a sense of continuity to the IOM's activities, so—in a perverse way—did the continuing efforts to define an appropriate relationship between the IOM and the NAS. Discussions on this subject continued throughout the first four years of IOM history. The desire to maintain a distinctive identity apart from the rest of the Academy helped unify IOM members. They examined communications from the NAS with the same suspicion and apprehension that eighteenth century American colonists reserved for letters from the King of England.

Philip Handler had a tendency to treat the IOM in an imperial manner. Indeed, John Hogness later claimed that Handler had never wanted the Institute to be created in the first place, but his other priorities prevented a major confrontation. These consisted of reorganizing the National Research Council and defining the relationship between the NAS and the National Academy of Engineering (NAE). Handler's relationship with Clarence Linder, the NAE president, was far worse than his relationship with John Hogness. In fact, both Linder and Handler used Hogness as a sounding board, each complaining about the other.[93] Despite the often prickly personalities involved, Handler persisted with his reorganization efforts. During 1971, he tried to enlist Hogness's help in reorganizing the National Research Council so that many IOM activities, particularly its ability to initiate studies, would be moved to the NRC. Hogness offered Handler little aid, telling him that IOM members would never go along with the changes. The IOM Executive Committee decided in November 1971 not to acquiesce in the reorganization plan. "IOM is a bastard organization. It does not fit properly anywhere," confessed Hogness to Handler.[94] For the next two years, Handler pursued the more sensitive negotiations with engineers. Then he told Hogness that his long years of bargaining with the NAE had produced an agreement that was adopted with enthusiasm by all of the parties involved, and it was a matter of "considerable urgency" to reach a similar agreement with the IOM.

Faced once again with a request that the IOM incorporate many of its activities into the NRC, the IOM Council responded by reasserting its basic principles. The Institute wished to retain its organizational structure, one that combined the characteristics of an honorary society and a working organization. The Council therefore rejected the notion that the IOM should become a purely honorary academy, with an "Assembly of Medicine" in the NRC as its "operating arm." The IOM also wanted to retain its multidisciplinary character, examining the "social science issues of medical care," as well as policies related to "health professional education and the biomedical and behavioral sciences basic to health." Although the Institute of Medicine agreed to be represented on the governing board of the National Research Council, it concurred in none of Handler's other suggestions, and for the moment at least, the NAS Council did not press the issue. In effect, the NAS accepted the fact that the IOM would differ in significant ways from the National Academy of Sciences and the National Academy of Engineering. Only one substantive change came about as the result of the 1973 reorganization effort: the authority to approve IOM projects was transferred from the NAS Council to the Governing Board of the NRC. "In essence," as John Hogness put it, "the present organizational situation is basically a good one and should be reaffirmed."[95]

Donald Fredrickson

When John Hogness reaffirmed the soundness of the IOM structure, he did so as someone who was about to leave the organization. Appearing before the IOM Council on May 10, 1973, he announced that he would assume the presidency of the University of Washington, the one job he felt he could not refuse, in the late spring of 1974. A search committee headed by Walsh McDermott moved quickly to select a successor. The committee chose Donald Fredrickson, who had the distinction of being a member of both the Institute of Medicine and the National Academy of Sciences.[96]

Two years younger than Hogness, Fredrickson had followed a different route in his career. He began his undergraduate studies at the University of Colorado, but the war interrupted them. After serving in the U.S. Army, he finished his undergraduate degree at the University of Michigan and remained in Ann Arbor for his medical degree. Between 1949 and 1953, he completed his training in hospitals and laboratories that were associated with Harvard. Beginning in 1953, he worked at the National Heart Institute, serving as its director from 1966 to 1968. In 1973, when the IOM search committee approached him, he was the director of intramural research at the National Heart and Lung Institute. Of all the IOM founders, his career resembled most closely that of James Shannon: the focus was on research, rather than on clinical practice.

Fredrickson had joined the IOM with the initial membership group in 1971, quickly becoming involved in it activities. In the fall of 1971, he received a phone call from Julius Comroe, who wanted another basic scientist to serve on the IOM Council and asked Fredrickson to run. Agreeing, Fredrickson became a Council member in 1972 and found it to be an engaging experience. "There was a rich mixture of the dialects and ethics operative in the world outside the laboratory walls," he recalled, that offered "an unparalleled view of the complex field of human health." When the search committee approached him in the fall of 1973, he was interested.

On November 20, 1973, Philip Handler invited Donald Fredrickson to his office. "The Council of the IOM thinks I ought to talk to you about John's job," Handler said. Fredrickson told him that the prospect of becoming president of the IOM was not unattractive but that he worried about how the NAS reorganization would affect the Institute. He defended the IOM practice of mixed membership, saying that "ecumenism" was important in such a complicated area as health. He announced that he would not consider the job unless the disagreements between the NAS and the IOM were settled. This provided a spur to Handler and Hogness and accounted in large measure for Handler's acquiescence to the IOM position. After some hesitation, Fredrickson accepted the offer of the IOM presidency on January 18, 1974.

The members' meeting in the spring of 1974 served as an inauguration of sorts for Donald Fredrickson. He spoke of bridging the worlds of science and medical practice, arguing that the IOM's commitment was to "lend the scientific method to the direction of a whole social movement." Noting that the IOM's success rested upon the "essence, not merely the appearance, of nonpartisan objectivity," he predicted that the same sort of satisfaction he had found in the laboratory could also be found at the Institute of Medicine.

However, Fredrickson faced a rude sort of culture shock. Perhaps the most jarring element was the need to visit the IOM's major funders, reassure them that the organization remained strong, and initiate the process of reapplying for support. As an official of NIH, Fredrickson had given away, not requested, money. Fund-raising in the federal government took the form of appealing to the Bureau of the Budget and Congress, a far different exercise than going to New York and meeting with foundation officials.

One exercise that the leadership of NIH and the IOM held in common was the need to provide sufficient office space for staff. The IOM staff was simply too large to be housed in IOM headquarters, nor could it secure enough space in an auxiliary building on Pennsylvania Avenue. Therefore, the IOM rented space in the Watergate office complex that had previously been occupied by the Democratic party. Indeed, Fredrickson's office turned out to be the same one used by Lawrence O'Brien, the very site of the Watergate break-in, which monopolized the headlines during this period surrounding President Nixon's resignation. As a consequence, IOM staffers often found their work interrupted by curious tourists seeking out the "stuff of history."

Fredrickson never had time to sink his teeth into the job as president of the IOM. Almost as soon as he arrived in the late spring of 1974, he began to receive phone calls from federal officials warning of dissension in the upper ranks at NIH. Some of these officials wanted to know if he would be willing to return to Bethesda as head of the National Institutes of Health. Early in January 1975, Fredrickson learned that he was on a short list of candidates for both Assistant Secretary of Health and director of NIH. He met with administration officials on January 9, 1975, a year to the day after he had received a phone call from Handler offering him the job as IOM president. On January 24, he told the IOM Council that he had been approached by the administration to serve as head of NIH and that he "would not refuse to consider this position" if it were offered to him. That morning, over breakfast, Adam Yarmolinsky had tried to dissuade Fredrickson from considering the position, arguing that he could do more good as head of the IOM. Fredrickson explained that as he had told Philip Handler, the NIH was "not a job; it's a cause." Realizing that something had to be done, the IOM Council moved to designate a vice chairman pro tem who would take over in Fredrickson's absence. It eventually selected Julius Richmond for the post.[97]

On April 19, 1975, HEW Secretary Caspar Weinberger announced the nomination of Donald Fredrickson as director of the National Institutes of Health. On May 5, Fredrickson addressed a letter to all members of the IOM telling them he had accepted President Ford's offer and that he would be leaving IOM in June. He said that the choice was difficult but he had chosen NIH because its needs seemed more critical. He realized that although changing presidents twice in four years created great potential for harm, he was confident that the IOM would "survive its president." He owed his confidence to his faith that Julius Richmond and Roger Bulger would preside effectively over the Institute during the transition period. He also took solace in the fact that the Robert Wood Johnson and Kellogg Foundations had renewed their core grants to the IOM and that the IOM would be able to maintain a budget of from $3.5 million to $4 million in fiscal year 1976.[98]

"In retrospect the time you have spent with us now seems to have been extraordinarily short," Handler wrote to Fredrickson.[99] Indeed, Fredrickson had only a few months to concentrate on the IOM. The rest of his short tenure was spent in negotiations with administration officials over the state of the National Institutes of Health. This left him with too little time to change the IOM's basic direction or to infuse the IOM with the funds that would make it truly self-sustaining. Of six major projects that the IOM had hoped to initiate during Fredrickson's tenure, only two had received any sort of outside funding.[100]

Conclusion

During the traditional August lull in 1975, Roger Bulger, who had been running the IOM on a daily basis since Fredrickson's departure, dictated a thoughtful letter to Larry Lewin, a Washington consultant who had worked on the IOM's costs of education study. The letter provided Lewin, who was hired to make recommendations on how the

IOM should be structured and managed, with an overview of the IOM's four-year history. It was an exceptionally candid look at the organization.

Bulger noted that the problems of the IOM began with its name. People mistook it for the National Institutes of Health. When the IOM did something newsworthy, its activities became confused with those of the National Academy of Sciences. As for the IOM's program, the organization had often been concerned with issues that were "very philosophic, sometimes ethical, often quite political, and frequently not as quantitative or analytical as may people would like." Still, most people knew of the organization only because of the large, data-collecting studies, such as the costs of education study. The problem was that it was possible just to collect data, without getting at underlying policy issues. Hogness and Fredrickson had both believed that much of the valuable activity at the IOM took place outside the formal studies. If nothing else, the IOM served to "broaden the horizons of each of its members." Many of its best conversations and activities took place in Council meetings and meetings of the Membership Committee. These discussions were themselves contributions to health policy.

Although the IOM had its headquarters in Washington, D.C., it was not a typical Washington organization. The IOM president did not make regular appearances on Capitol Hill or serve as a sort of "health policy guru." In this regard, the IOM differed from the American Association of Medical Colleges, whose president, John Cooper, had become a major player on the national scene. Some people thought that the IOM should try to play a more visible role in the policy process. Part of what inhibited the IOM was the inability to distribute the results of its studies. Bulger thought it sad that the IOM could not provide copies of its costs of education studies to every medical school in the country. Nor did the IOM spend much on publicity, depending on the "kindness of strangers" to let the world know of its accomplishments.

As a way to overcome some of these problems, John Hogness had "utilized the development of a staff as a combination of acquiring expertise and analytic abilities with extending the points of contact for the Institute throughout the health establishment." The members of the staff complemented one another. Bulger had experience with the allied health professions, Karl Yordy with the government, and Ruth Hanft with health services researchers. The IOM used its visiting scholars and residents to fill in the gaps. Every such request was scrutinized with an eye toward the potential "contribution to the general environment of the Institute and its possible contribution to the staff." The result, according to Bulger, was a "reasonable level of staffing." "We have a very wholesome interdisciplinary thing going," he added, "such that I am comfortable as a health professional with the whole range of staff people."

Bulger concluded his long letter by saying that "as we have developed more of track record and more people have taken notice of us, we have also attracted our share of jealousies and legitimate concerns. It is important that in this city of Washington, those people who might look to us for advice and who might be able to use us should think of us as competent and able to deal with the problems."[101] Hogness and Fredrickson had made a start in this direction, in particular by raising funds from foundations and establishing the basic routines of IOM governance and IOM studies. Still, as Bulger had noted, the organization remained invisible to mainstream Washington and the general public. By the time Jimmy Carter became president in 1977, this condition would change.

Notes

1. Philip Handler to Dr. Ward Darley, January 20, 1971, Institute of Medicine, 1971, General, National Academy of Sciences (NAS) Records.

2. Roger Bulger to Larry Lewin, August 6, 1975, Yordy Files, Accession 91-051, Institute of Medicine (IOM) Records, NAS Archives.

3. Oral interview with Robert Glaser, Menlo Park, Calif., November 1, 1997.

4. Meeting of Executive Committee, December 17, 1970, Institute of Medicine, NAS Records.

5. Minutes of Initial Membership Committee, January 28, 1971, Robert Glaser Papers, Menlo Park, Calif.

6. Initial Membership Committee, March 10, 1971; Robert J. Glaser to Philip Handler, April 16, 1971; and Inger Herman to Handler, June 21, 1971, all in IOM Records, NAS Archives; Glaser to Julius Comroe, April 5, 1971, Glaser Papers.

7. Raymond Waggoner, immediate past President of the American Psychiatric Association, to Philip Handler, June 15, 1970; Isidore Greenberg, Director of Graduate Studies, Brooklyn College of Pharmacy, to David Hamburg, June 15, 1970; Eric Comstock, Executive Director, American Academy of Clinical Toxicology, to Handler, July 18, 1970; R. L. Kitchell, Dean of the College of Veterinary Medicine, Iowa State University, to Handler, July 21, 1970; and Joseph Goodgood, Professor of Rehabilitation Medicine, New York University, to Handler, August 13, 1970, all in IOM Membership, 1970, NAS Records.

8. Robert Glaser to Philip Handler, April 16, 1971, IOM Membership, 1971, NAS Records.

9. Robert Glaser to Institute of Medicine, April 27, 1971, Glaser Papers.

10. Robert Glaser to Philip Handler, April 16, 1971, and Inger Herman to Handler, June 21, 1971, IOM Membership, 1971, NAS Records.

11. Robert J. Glaser to Philip Handler, January 20, 1971; Handler to Dr. John Hogness, March 24, 1971; and NAS Press Release, March 30, 1971, all in Institute of Medicine, Administrative, President, 1971, NAS Records.

12. Harold M. Schmeck, Jr., "President of the Institute of Medicine: John Rusten Hogness," *New York Times,* March 30, 1971.

13. John R. Hogness to Wallace K. Waterfall, June 20, 1990, Waterfall Materials, Institute of Medicine; Oral interview with John Hogness, Mazama, Wash., October 1, 1997.

14. Curriculum Vitae for John R. Hogness, NAS Records.

15. Clipping from *Medical World News,* May 7, 1971, in Institute of Medicine, Administrative, President, 1971, NAS Records.

16. *Ibid.*

17. "Institute's Goal: 'Fresh, Unbiased' Look at Medicine," *American Medical News,* September 20, 1971, p. 13.

18. "Hell of a Show Promised by Hogness for Institute of Medicine: Organizational Work Finished, Funds Promised, He Tells Meeting," *Drug Research Reports,* May 24, 1972, Irvine Page Papers, National Library of Medicine, Bethesda, Md.

19. IOM Executive Committee Meeting, Minutes, December 21, 1971, NAS Records.

20. IOM Executive Committee Meeting, Minutes, September 16, 1971, IOM General, 1971, NAS Records.

21. Oral interview with Roger Bulger, June 11, 1997, Baltimore, Md.

22. IOM Executive Committee Meeting, Minutes, December 21, 1971, NAS Records.

23. IOM Executive Committee Meeting, Minutes, September 16, 1971, IOM General, 1971, NAS Records.

24. IOM Executive Committee Meeting, Minutes, June 21, 1971, NAS Records.

25. John Walsh, "Institute of Medicine: Broad-Spectrum Prescription," *Science* 74 (November 26, 1971), pp. 929–933.

26. John Hogness to Wallace Waterfall, June 20, 1990, Waterfall Materials.

27. IOM Executive Committee Meeting, Minutes, March 11, 1970, NAS Records.

28. IOM Executive Committee Meeting, Minutes, June 10, 1971, NAS Records.

29. IOM Executive Committee Meeting, Minutes, September 16, 1971, IOM General, 1971, NAS Records.

30. "Hogness Testified Against National Cancer Authority at Executive Committee's Direction," *Drug Research Reports* 14 (October 27, 1971), p. 6, in Page Papers; Stephen P. Strickland, *Politics, Science, and Dread Disease: A Short History of Medical Research Policy* (Cambridge, Mass.: Harvard University Press, 1972), p. 287; see also James T. Patterson, *The Dread*

Disease: Cancer and Modern American Culture (Cambridge, Mass.: Harvard University Press, 1987).

31. IOM Council Meeting, Minutes, September 14, 1972, IOM Records, NAS Archives.

32. "The National Cancer Program Plan: A Report of the Ad Hoc Review Committee of the Institute of Medicine, National Academy of Sciences," December 15, 1972, in Materials for IOM Council Meeting, January 18, 1973, IOM Records.

33. John Hogness to Philip Handler, November 29, 1972, in Materials for IOM Council Meeting, January 18, 1973, IOM Records.

34. Philip Handler to John Hogness, November 30, 1972, in Materials for IOM Council Meeting, January 18, 1973, IOM Records.

35. John Hogness to Philip Handler, December 4, 1972, in Materials for IOM Council Meeting, 1972, IOM Records.

36. IOM Program Committee Meeting, Minutes, May 1973, Accession 81-006, IOM Records.

37. See Paul J. Sanazaro, "Federal Health Services R&D Under the Auspices of the National Center for Health Services Research and Development," in Sanazaro and E. Evelyn Flook, eds., *Health Services Research and R&D in Perspective* (Ann Arbor, Mich.: Health Services Press, 1973), pp. 150–183.

38. Action Memorandum for Executive Committee of the Council of NAS, November 13, 1971, in Accession 86-064-04, IOM Records.

39. Philip Handler to Robert Glaser, March 16, 1971, and Glaser to Handler, March 22, 1971, both in Glaser Papers.

40. Leon White to John Hogness, June 18, 1971, and Hogness to White, June 23, 1971, both in Universal Entitlement Study Files, IOM Records.

41. IOM Executive Committee Meeting, Minutes, September 16, 1971, IOM General, 1971, NAS Records.

42. Action Memo for Executive Committee of the NAS, November 13, 1971.

43. IOM Program Committee Meeting, Minutes, December 1971, IOM Program Committee Files, NAS Records.

44. David Mechanic to John Hogness, December 17, 1971, Entitlement Study Files, IOM Records.

45. "Minutes of Meeting of February 5, 1972," Boston, and Irving Lewis to John Hogness, February 9, 1972, both in Accession 86-064-04, Entitlement Study Files, IOM Records.

46. Martin Feldstein to John Hogness, April 3, 1972, Accession 86-064-04, Entitlement Study Files, IOM Records.

47. Leon White to John Hogness, September 16, 1971, Accession 86-064-04, Entitlement Study Files, IOM Records.

48. John Hogness to Martin Feldstein, April 14, 1972, Accession 86-064-04, Entitlement Study Files, IOM Records.

49. Barbara J. Culliton, "Institute of Medicine: Taking on Study of Cost of Medical Education," *Science* 176 (June 1972), pp. 998–999; David Kessner to Panel on Health Services, March 20, 1969, Accession 95-050, IOM Records;

Helen C. Chase, David M. Kessner, et al., eds., "Infant Death: An Analysis by Maternal Risk and Health Care: Contrasts in Health Status, Volume 1," 1973, Accession 95-050, IOM Records.

50. See Minutes of IOM Meeting, January 18, 1973, IOM Records.

51. M. Mitchell-Bateman, M.D., to John Hogness, March 14, 1974, Records of the Study on the Health Effects of Legal Abortion, Accession 81-008, IOM Records (hereafter, Abortion Study Files).

52. See, for example, Donald Fredrickson to Dr. Leslie W. Dunbar, Executive Director, Field Foundation, December 19, 1974, Abortion Study Files, IOM Records.

53. See, for example, Martha Blaxall to Steering Committee, August 29, 1974, Abortion Study Files, IOM Records.

54. Institute of Medicine, "Report of a Study, Legalized Abortion and the Public Health," May 1975, Abortion Study Files, IOM Records.

55. Jason Douglas, Director, Center for the Study of Moral Order, to Donald Fredrickson, June 6, 1975, Abortion Study Files, IOM Records.

56. Roger J. Bulger to Mr. S. G. Landfather, Executive Director, Sunnen Foundation, September 29, 1975, Abortion Study Files, IOM Records.

57. Materials for Action, January 18, 1972, IOM Records; Institute of Medicine, *Disease by Disease Toward National Health Insurance* (Washington, D.C.: National Academy of Sciences, 1973), p. 10.

58. John Hogness to Paul Ward, November 21, 1972, Statement on Health Maintenance Organization Files, Accession 95-021, IOM Papers, (hereafter, HMO Files). On the HMO Act of 1973, see Lawrence D. Brown, *Politics and Health Care Organization: HMOs as Federal Policy* (Washington, D.C.: Brookings Institution, 1983).

59. Institute of Medicine, "Health Maintenance Organizations: Toward a Fair Market Test: A Policy Statement by a Committee of the Institute of Medicine," May 1974, HMO Files, IOM Records.

60. Anne Somers to J. F. Volker, D.D.S., Ph.D., January 14, 1974, and Harold F. Newman, M.D., to Karl Yordy, April 19, 1974, both in HMO Files, IOM Records.

61. Merlin K. DuVal to Dr. John R. Hogness, November 17, 1971, Materials for IOM Council Meeting, IOM Records.

62. See "Interim Report to Congress: Costs of Education of Health Professionals," March 15, 1973, Accession 86-064-03, Files of the Study on Costs of Educating Health Professionals, IOM Records (hereafter, Costs of Education Files).

63. IOM Council Meeting, Minutes, March 16, 1972, IOM Records; John Hogness oral interview; Oral interview with Ruth Hanft, June 10, 1997, Washington, D.C.; Testimony of Roger J. Bulger before Subcommittee on Health and Environment of the House Interstate and Foreign Commerce Committee, May 20, 1974, Costs of Education Files, IOM Records.

64. See John Ingle, Senior Staff Officer, to Harold Helms, Dean, School of Pharmacy, University of Colorado, March 27, 1973; Ruth S. Hanft, "Notes for a Speech to Association of Health Professional Schools," September 19, 1972;

Memorandum by Eli Ginzberg, September 11, 1972, and other materials, all in Costs of Education Files, IOM Records.

65. IOM Steering Committee Meeting, Minutes, September 18–19, 1973, Costs of Education Files, IOM Records.

66. "Costs of Education in the Health Professions, Summary," January 1974, and "Costs of Education in the Health Professions, Parts I and II," January 1974, both in Costs of Education Files, IOM Records.

67. Ruth Hanft, "Testimony of the Institute of Medicine, National Academy of Sciences before the Subcommittee on Health and Environment of the House Interstate and Foreign Commerce Committee," May 20, 1974, Costs of Education Files, IOM Records.

68. IOM Council Meeting, Minutes, September 19, 1974, IOM Records.

69. John Hogness to Richard Raring, September 20, 1971, NAS Records.

70. "Annual Report for the Year Ending June 30, 1972," p. 6, in Materials for IOM Council Meeting, January 18, 1973, IOM Records.

71. IOM Council Meeting, Minutes, July 20, 1972, IOM Records.

72. "Program Evaluation Groups," in IOM Council Meeting, Minutes, March 16, 1972, IOM Records; James A. Shannon to Roger Bulger, March 28, 1972; Philip Handler to Shannon, April 4, 1972; and Bulger to Shannon, April 3, 1972, all in Yordy Files, Accession 94-111, IOM Records.

73. "The Development of the Program of the Institute of Medicine," draft November 8, 1972, in agenda items for IOM Council Meeting, November 16, 1972, IOM Records.

74. John Hogness to Wallace Waterfall, June 20, 1990, IOM Records.

75. Figures come from "Draft Annual Report for the Year Ending June 30, 1972," in Materials for IOM Council Meeting, January 18, 1973, IOM Records.

76. *Ibid.*

77. John Hogness to Gustav Lienhard, President, Robert Wood Johnson Foundation, December 16, 1971; Hogness to David E. Rogers, January 26, 1972; Walter J. Unger to Donald S. Fredrickson, January 10, 1975; and Rogers to Philip Handler, June 2, 1972, all in Funding Files, Accession 91-005, IOM Records.

78. Roger Bulger to David E. Rogers, September 13, 1972, Accession 91-045, IOM Records.

79. John Hogness to David Rogers, June 20, 1973, Funding Files 1972–1981, Accession 90-074, IOM Records.

80. John Hogness to Frederick W. Featherstone, M.D., Program Director, W. K. Kellogg Foundation, February 28, 1972; Hogness to Featherstone, April 26, 1972; Featherstone to John Coleman, Executive Officer, NAS, June 28, 1972; News Release, "Institute of Medicine Receives Kellogg Foundation Grant," July 31, 1972; and Hogness to Featherstone, August 3, 1973, all in Funding Files, Accession 91-045, IOM Records.

81. Robert J. Glaser to John Hogness, March 22, 1972; Hogness to George Taber, June 7, 1972; Richard P. Mellon to Hogness, December 15, 1972; and Hogness to George Taber, September 29, 1973, all in Funding Files, Accession 91-045, IOM Records.

82. John Hogness to Quigg Newton, February 5, 1974; Roger Bulger to Aaron Rosenthal, NAS Comptroller, April 9, 1974; Walter Unger to David Hamburg, February 1, 1977; Hogness to Nathan Pusey, January 7, 1974; "Proposal to Andrew W. Mellon Foundation," November 18, 1974; and Pusey to Philip Handler, December 20, 1974, all in Funding Files, Accession 91-045, IOM Records.

83. Memorandum for the Files, "New York Trip Report," October 29, 1975, Funding Files, Accession 91-045, IOM Records.

84. Adam Yarmolinsky to McGeorge Bundy, January 21, 1974; Bundy to Yarmolinsky, January 31, 1974; Bundy to Roger Bulger, March 22, 1974; and Harold Howe to Walter Unger, November 21, 1974, all in IOM Funding Files, Accession 91-045, IOM Records.

85. William Danforth, Chancellor, Washington University to Donald Frederickson, April 7, 1975, Funding Files, Accession 91-045, IOM Records.

86. John Hogness to Philip Handler, June 1, 1972, and Irvine Page to Hogness, August 22, 1972, both in IOM Membership, 1972, NAS Records.

87. John Hogness to Irvine Page, September 19, 1972, IOM Membership, 1972, NAS Records; Philip Handler to Hogness, June 13, 1973, and David Mechanic to Roger Bulger, both in Materials for IOM Council Meeting, July 1974, IOM Records; IOM Council Meeting, Minutes, March 20, 1975, IOM Records.

88. Institute of Medicine, *Annual Report,* year ended June 30, 1975 (Washington, D.C.: National Academy of Sciences, 1976) p. 10.

89. IOM Council Meeting, Minutes, March 15, 1973, IOM Records.

90. David Rogers to John Hogness, March 29, 1973, Funding Files, Accession 91-045, IOM Records.

91. Robert Marston to David Rogers, June 27, 1974, Funding Files, IOM Records; Barbara J. Culliton, "Johnson Health Policy Fellows: Joining the Scientific and Political," *Science,* September 19, 1973, pp. 977–980.

92. David Rogers to Philip Handler, April 5, 1976, Funding Files, Accession 91-045, IOM Records.

93. John Hogness to Wallace Waterfall, June 20, 1990, IOM Papers.

94. Executive Committee Meeting, Minutes, November 18, 1971, and December 21, 1971; John Hogness to Philip Handler, August 18, 1971; Hogness to Handler, September 22, 1971; and Hogness to Handler, November 30, 1971, all in Yordy Files, Accession 91-051, IOM Records.

95. Philip Handler to John Hogness, November 7, 1973, Yordy Files, Accession 91-051, IOM Records; Special Meeting of the IOM Council, Minutes, December 19, 1973, attached to Hogness to Handler, January 4, 1974, IOM Records; and Handler to Hogness, February 13, 1974, Yordy Files, Accession 91-051, IOM Records.

96. Much of what follows is based on Donald S. Fredrickson, "I.O.M.: Mémoires à l'Accouchement," an essay on his IOM experiences that he prepared for the Institute's 20th anniversary, in the IOM Records.

97. Memorandum for the Record, January 24, 1973, IOM Council Minutes, IOM Records.

98. Donald Fredrickson to Members, Institute of Medicine, May 5, 1975, Attachment to IOM Council Meeting, Minutes, May 15, 1975, IOM Records.

99. Philip Handler to Donald Fredrickson, June 25, 1975, IOM-1975 File, NAS Records.

100. Walter J. Unger to Donald Fredrickson, Julius Richmond, and Nathan Start, May 12, 1975, attachment to IOM Council Meeting, Minutes, May 15, 1971, IOM Records.

101. Roger Bulger to Larry Lewin, August 6, 1975, Yordy Files, Accession 91-051, IOM Records.

3

The Hamburg Era

On May 3, 1977, President Jimmy Carter read an article in the *Washington Post* and jotted down a note for Joseph Califano, his Secretary of Health, Education, and Welfare (HEW). The President wanted Califano to read the story on an Institute of Medicine (IOM) study of computed tomography (CT) scanners. The thrust of the study was that hospitals and physicians should not overuse the beneficial yet costly new technology. Only after the local health planners approved should a new CT scanner be installed; these local planners should make sure that each scanner operated at maximum efficiency, performing a minimum of 2,500 tests a year. Hospitals, not private doctors' offices, should be the setting for scanners. "Let's take similar action—stronger if possible—and include other devices as well," the President urged his Secretary of Health, Education, and Welfare.[1]

If the Institute of Medicine wanted evidence of its influence, the note from Carter to Califano provided it. In 1977, the IOM knew that the President of the United States and his chief minister for health both took an interest in its activities. The moment marked the realization of Walsh McDermott's aspirations by proving that the IOM could have an enduring effect on the nation's health policy. It also demonstrated the dramatic results of David Hamburg's management. Hamburg, who had taken over from Donald Fredrickson in the fall of 1975, provided the IOM with five years of inspirational leadership.

Despite these considerable achievements in the Hamburg era, the embrace of the Carter administration failed to heal all of the IOM's maladies. Even at the end of five years of steady activity, old problems continued to nag at the organization. These included the lack of a secure financial base, troubled relations with the National Academy of Sciences (NAS), and an often less than clear focus for the IOM's activities.

David Hamburg Arrives

The departure of Donald Fredrickson put a strain on the IOM, which had gone through two presidents in less than five years. Most of the effort to pick up the pieces fell to Julius Richmond, the distinguished child psychiatrist who acted as vice chairman pro tem of the IOM Council. Much as Robert J. Glaser had filled the void between Walsh McDermott and John Hogness, so Richmond, ably assisted by Roger Bulger, kept things together in the spring and summer of 1975. He quickly constituted the Executive Committee of the Council as a search committee and set it to work finding a new IOM president.

The fact that the IOM had searched for a president less than two years earlier facilitated the process. The same short list that had been developed in the summer of 1973 could also be used in the spring of 1975. Although Richmond encouraged members to submit nominations and generated a list of more than 60 names, he and his committee concentrated from the first on a few, select candidates. On June 19, the IOM Council culled through the names and designated David Hamburg as its first choice. The NAS Council moved quickly to confirm the selection, although Philip Handler objected to being given only one name to consider. Richmond explained that although the Council had other candidates in mind, it had not "ranked them as clearly as we had Dr. Hamburg." Handler acquiesced, in part because Hamburg had been the second choice less than two years before and would, in effect, be filling Fredrickson's unexpired term. On June 26, 1975, Handler wrote to Hamburg and offered him the job.[2]

David Hamburg's acceptance of the offer depended on a set of circumstances as bizarre as the IOM had ever encountered. Because Hamburg was in Dar es Salaam, Handler's letter reached him very slowly. Hamburg had gone there to negotiate the release of four Stanford students who had been kidnapped by Zairian rebels. On July 19, 1975, Hamburg reported to Handler that three of the four students had been freed without harm, "but the negotiating process for the fourth is very difficult." In the meantime, he and his wife were discussing Handler's offer.[3] The kidnapping incident delayed the recruiting process and meant that Roger Bulger effectively ran the IOM from the beginning of July until the beginning of November. Formal announcement of Hamburg's appointment came only at the end of October.

Born in 1925, David Alan Hamburg spent his childhood and received much of his education in Indiana, graduating from the University of Indiana Medical School in 1947. After taking three years off for military service, he went to Michael J. Reese Hospital, where he had taken much of his postgraduate training, and worked at the Institute for Psychosomatic and Psychiatric Research and Training. In 1958, he began a three-year stint as chief of the adult psychiatry branch at the National Institute of Mental Health in Bethesda. Then, in 1961, he took over as the chairman of Stanford's Department of Psychiatry. In the summer of 1975, he held the appointment as Reed-Hodgson Professor of Human Biology in Stanford's medical school. In his research, Hamburg analyzed the links between biology and behavior, examining such things as the behavioral, endocrine, and genetic aspects of stress and the biologic basis for aggressive behavior. He had more than 93 scientific papers to his credit. The appointment of Hamburg continued the IOM tradition of selecting distinguished researchers and administrators from the field of academic medicine as president.[4]

On November 6, 1975, David Hamburg delivered his inaugural remarks to IOM members who attended the annual meeting. He was far from a stranger in IOM circles, having been selected as an initial member in 1971, served on the Council from 1971 to 1974, and chaired the Program Committee between 1972 and 1973. In his inaugural remarks, he described what he considered the key features of the IOM. The organization made a "serious, thoughtful attempt to face" difficult issues and to do so in a way that cut "across traditional specialties and perspectives." Unlike other organizations, the IOM held "no over-riding doctrine, no party line, no cow too sacred to be examined," and its views reflected "deeper analysis and reflections" than those of others in the health field. After listing the organization's assets, Hamburg pondered its liabilities. He wondered how the IOM could coax sufficient time from its busy members to examine the key issues, how a Washington-based staff could relate to a geographically dispersed membership, how the IOM could preserve its independence and be of use both to the government and to private institutions, and how the IOM could achieve cross-specialty collaboration "to tackle the policy problems of health care, prevention, education and the science base underlying it all." These problems, Hamburg observed, deserved "thoughtful attention. And I intend to get it—by quiet, respectful inquiry if possible; by relentless harassment if necessary."[5]

Speaking to the members, Hamburg revealed an almost pastoral style. Almost alone among the previous IOM presidents, he spoke and wrote with an inspirational eloquence with which he exhorted the members to contribute to the organization. At the same time, he realized that he and the organization faced many problems, not the least of which were his own logistical difficulties. He told the IOM

Council, for example, that he would not be able to move to Washington on a full-time basis until June 1976. Until then, he would take up the IOM presidency to the "maximum extent that circumstances would permit."[6]

Ruth Hanft and the Social Security Studies

Hamburg would also have to build up his own team. At the end of 1975, Adam Yarmolinsky ended his term on the IOM Council, severing one of the last links between the Institute's founders and the inner circles of IOM policymaking. In the case of Yarmolinsky, the Council thought so highly of his services that it took steps to preserve a role for him, designating him as a special counselor to the IOM. He continued to provide the organization with legal and practical advice until, accepting a position in the Carter administration, he suspended his close IOM ties for the duration of his government service. Another key departure, that of Roger Bulger, took place early in 1976. He became chancellor of the University of Massachusetts at Worcester and dean of the University of Massachusetts Medical School. No doubt part of his appeal to the university was the fact that he had served so effectively as interim head of the IOM and had gotten to know so many leading figures in academic medicine through his service as the IOM's executive officer. Still another departure that took place in 1976 involved Ruth Hanft.

With Hanft went the large data-gathering studies. As David Hamburg took hold of the IOM, another such study reached completion. This second study followed from the first one, which had concerned the costs of medical education. After the IOM had submitted the earlier study, Congress asked it to consider the related problem of how the Medicare and Medicaid programs should pay physicians in teaching hospitals for their treatment of elderly (Medicare) or indigent (Medicaid) patients. Congress also wanted to know how much federal money went into the support of foreign medical school graduates and how Medicare reimbursement could be used to avoid gluts of physicians in some areas and shortages in others. Finally, the study request asked the IOM to consider how Medicare reimbursements might be structured so as to encourage a greater number of physicians to enter primary care fields.

Adam Yarmolinsky chaired the steering committee, and Ruth Hanft directed a large staff, which eventually grew to 45 people, for what became a very elaborate effort. This second large data-gathering study raised many of the same problems as the first. The Medicare law, passed in 1965, made a distinction between payments for medical services provided by physicians and payments for medical services provided by hospitals. In the case of teaching hospitals, however, it was difficult to separate physician and hospital services. Medical students, interns, and residents all treated patients, yet these practitioners often received salaries from hospitals or medical schools as part of the hospital staff. Furthermore, it was difficult to separate the costs of treatment from the costs of education. To address these problems, the staff designed a survey that was sent to 1,400 teaching hospitals; it made site visits to 96 teaching hospitals, 15 medical schools, and 2 osteopathic schools and ascertained how the more than 100 intermediaries and carriers who administered Medicare on behalf of the Social Security Administration defined teaching physicians and teaching hospitals.

The collaboration between Yarmolinsky's steering committee and Hanft's staff often grew strained. In the final version of the report, Yarmolinsky attributed classical virtues to Hanft and her staff. They were "heroic," "stoic," and "Socratic in [their] dialogue with the steering committee." At the same time, staff and steering committee faced different and often conflicting tasks. The staff wanted to collect as much data as accurately as possible and be responsive to Congress. Although the steering committee shared this goal, it also sought to draw larger policy implications from the data. The process of combining these different outlooks and missions into a coherent document, on a subject technical enough to require a seven-page glossary just to make the report accessible to medical experts, proved difficult. In the end, the IOM issued 99 pages of findings and recommendations and many more pages of what were described as resource papers.

Despite Yarmolinsky's insistence on clarity, the study's recommendations were so inward and technical as to be beyond most people's comprehension. For example, the group recommended that "under the cost payment regulations issued under Section 15, Public Law 93-233, inclusion of payment of the imputed value of volunteer services should be continued; under the same cost regulations, the ceiling of $30,000 on the imputed value of a volunteer teaching physician's services should be changed to the average salary for full-time physicians in the area or the VA [Veterans Administration] compensation for full-time physicians if an area average is unavailable." In general, however, the group recommended that an interim solution to the problem of Medicare reimbursement in teaching hospitals, adopted by Congress in 1972, be discontinued and that other payment methods, which the group specified, might be appropriate. In addition, the group came out in favor of providing more funds for ambulatory care services so that teaching hospitals could more easily support primary care training programs. Going even further, the study recommended elimination of the immigration law incentives for importing foreign physicians. According to Yarmolinsky, the most controversial recommendation of all was one that would permit the Secretary of HEW to withhold Medicare and Medicaid funds from residency programs in specialties that a "permanent, quasi-independent physician manpower commission" determined to be in excess supply.[7]

The IOM held a press conference to release the report. Questions dealt with the recommendations favoring ambulatory care services because, as Ruth Hanft conceded, the proposals in this area were easier to understand than the others. Hanft also briefed congressional staffers on the report; they tended to share some of her dissatisfaction with the IOM study process. The report recommendations, the congressional staff members believed, confused the interests of the medical schools with the public interest, even if the study itself had produced valuable data. Expressing unhappiness with the composition of the steering panel, the congressional staff members thought it was weighted, like IOM membership itself, in favor of academic medical centers.[8]

Most IOM members never had a chance to read the Medicare reimbursement report. Because it was large, bulky, and expensive to mail, it went only to members of the steering committee and the Council. When David Hamburg learned about this situation, it made him think more about how to engage members in the work of the IOM. His experiences with the Medicare reimbursement study also caused Hamburg to consider what role, if any, large data-gathering studies should play in the IOM program. Ruth Hanft, for her part, pushed Hamburg to accept other, technical assignments from Congress that did not require issuing recommendations and could be done by staff alone. Concerned about the separation of staff and membership, Hamburg and the IOM Council grew wary of Hanft's requests, and she eventually left the IOM staff.[9]

The Institute Recharts Its Course

Although Hamburg wished to encourage member participation, he realized how important it was to maintain good relations with the congressional staffs who were interested in using the IOM to create and maintain data bases. The mid-1970s marked a period of resurgence in congressional activism that followed the Watergate investigations. In this era, the bipartisan spirit that had prevailed between the Republican President and the Democratic Congress began to break down, and Congress began to insist on developing its own analytical capacity rather than accepting the advice of executive agencies. Congressional subcommittees, staffs, and agencies such as the Congressional Budget Office and the Congressional Research Service all began to grow. Health, particularly the costs of health care, was a major preoccupation of congressional and executive branch policymakers in this era of concern for rising prices and stagnant wages. Although Hamburg wanted to make the IOM relevant to this new policymaking structure, he hesitated to do so in a way that would turn the IOM into a staff-driven Washington think tank or consulting firm.

Like his predecessors, Hamburg grappled with the issue of the IOM's identity during his first year in office. He initiated a major review of the IOM's progress that lasted from the fall of 1975 through the fall of 1976. The review, extending to all levels of the IOM's operation—from the staff, to the IOM Council, to the IOM membership—provided Hamburg with a means of learning about the organization, forging a consensus as to its future direction, and reshaping the organization to reflect this consensus. In a more personal sense, it enabled Hamburg to fill in the awkward period during which he commuted to Washington from California.

The process began at the staff level with regular staff discussions and with the work that Roger Bulger had commissioned Larry Lewin to do in the summer of 1975. At Bulger's suggestion, Lewin, a noted management consultant in the health care field, questioned a wide range of people on a broad range of topics about the IOM and presented his results in the form of a staff seminar that took place on March 19, 1976.[10] After he told the staff some of what people had said to him about the IOM, he led a structured discussion designed to expose problems and suggest possible solutions.

As Lewin's report made clear, people in the Washington community held differing opinions of the IOM's capabilities. Jay Constantine, a plain-speaking member of the Senate Finance Committee staff, said that the IOM should perform "neutral" studies for Congress, unlike the costs of education study on which the IOM had "whored" by ignoring the evidence and coming out in favor of capitation. Stuart Altman of the Department of Health, Education, and Welfare thought that the IOM was needed, but not for large data-gathering studies because others could do them better. Jim Mongan, a colleague of Constantine's, agreed and said that the federal government should develop its own analytic capacity, through agencies such as the congressional Office of Technology Assessment, rather than depend on the IOM. The weight of opinion, therefore, seemed to be against the IOM's doing large data-gathering studies. As for what the IOM should do, one public health official believed that conferences should be its main activity; others urged the IOM to do quick-response studies, perhaps without using a steering committee.

The discussion then turned to the management of the IOM. One lingering problem was an undertone of antagonism between staff members assigned to specific projects and members of the permanent core staff. Project staff perceived that the core staff all too readily identified with the IOM membership rather than with their fellow staff members. The problems were particularly acute on steering committees. Many staff members felt that such committees were a "resource to be managed," rather than the ultimate arbiters of a particular study or project. Listening to the discussion, David Hamburg realized that he would have to clarify the expectations that project staffs and steering committees had of one another.

Although the discussion was diffuse, it did produce four tentative conclusions. First, the IOM, the staff now believed, should work on some big issues but not at the expense of monopolizing IOM money or staff talent. Second, the IOM staff should develop the ability to perform policy analysis on a quick-response basis. Third, the organization should work hard to alter the "wrong but strong" perception that the IOM was dominated by officials from academic health centers. Fourth, the IOM should use the broad-based interests of its members to expand the Institute from a Washington-oriented operation to a national operation. More than anything else, however, the discussions demonstrated that no one had a firm fix on the IOM's mission. Adjectives such as "rudderless," "drifting," and "floundering" came up in Lewin's discussion with government and foundation officials. According to one foundation officer, "the IOM had not yet decided what its mission should be." Furthermore, the IOM did not function well enough "to free its president to chart a course and to harness available resources to pursue that course."

Staff tried to follow up on the Lewin session by posing their own questions about IOM's organization and management. In effect, staff members prepared a "wish list" for Hamburg to consider. High on the list was the desire to clarify the relationship between the staff and the steering committees. Staff wanted steering committee members to receive a document that, among other things, explained the staff role, exhorted committee members to meet deadlines, educated the committee on basic research methodology, and admonished steering committee members to attend meetings. The staff also pressed for studies that its members could do on their own, with only an advisory, not a steering, committee to assist them.[11]

Hamburg made few commitments and turned next for advice from the IOM Council and from IOM members. Council member Dorothy Rice, director of the National Center for Health Statistics, gave one of the most thoughtful replies. She wrote that most members felt the IOM had not lived up to its great potential, spending too much time on large studies that did not address the nation's most important health problems. Rice told Hamburg that the IOM had to decide how much of its efforts should be devoted to "major studies, policy statements, background papers and conferences and seminars." She then asked how the IOM should operate. Should it be like Brookings, which turned a talented staff loose on important problems; the Committee on Economic Development, which issued policy statements that reflected the views of enlightened businesspeople on policy issues; or the Urban Institute, which performed contract research for the federal government? As the IOM grappled with this issue, Rice cautioned that Hamburg would have to consider the important issue of staff morale. "To keep good staff," she advised, "you need to provide a mechanism for close identifications of the individual and the research product." This led her to the issue of steering committees. Was their purpose to act as an advisory group or "to provide actual directions to staff in the conduct of the study"? When studies were completed, it often took a great deal of time to disseminate the results. Rice said that the delays relating to project approval that were built into the system required examination, as did the question of how the membership might be involved in the entire process.[12]

A major motif of the members' responses to Hamburg was that the IOM was not the organization they wanted it to be. Ernest Saward, former director of the Permanente Clinic in Portland, Oregon, and an original member of the Board on Medicine, told Hamburg that he was frustrated with the IOM. "It was to be a working group, not an honorific one," he explained, and "while there was to be staff, it hardly was imagined that the staff did all the work and the members were more or less judicial." Saward complained that unless one were a member of the Council, it was hard to keep up with IOM activities. The Board's original vision for the IOM "seems very distant to us now."[13]

Robert Petersdorf, the head of the Department of Medicine at the University of Washington, agreed with Saward that "the Institute is not fulfilling the high hopes which many had for it," but he disagreed on the nature of the failure. In Petersdorf's opinion, the IOM was wanting because it had failed to act as Irvine Page had hoped it would. "The IOM has not really become the collective spokesman for medicine," he told Hamburg. The problem lay in the fact that social scientists had taken over the IOM, in part because they had worked harder at it. Although Petersdorf conceded that social scientists were necessary, he worried that many had made their reputations by "taking anti-medicine stands. In fact, a number of people have begun to call the IOM the 'Institute of Anti-Medicine.' " It was up to Hamburg to set things right. "No organization," Petersdorf advised, could prosper "unless there is strong direction from the top and in a relatively continuous fashion."[14]

Before Hamburg provided this direction, he wanted a mandate from the Council and the members. Toward this end, he scheduled a series of retreats for the IOM Council that took place in July 1976. Julius Richmond presided over the first one in which five IOM Council members held a long and unstructured discussion of IOM priorities, deciding that the IOM should devote more of its time to "short-term, quick-response kinds of activities" that allowed the organization to react to public policy developments.[15] Hamburg seemed to affirm this point at a meeting of the Program Committee that took place a week later. He said that the Institute would not be doing so many "responsive" studies in the style of the Medicare–Medicaid reimbursement project. Instead, it would turn more toward private foundations and corporations to support smaller studies.[16]

At the end of July, Council members held a two-day retreat at Woods Hole, Massachusetts, at which they contemplated the IOM's future. Setting the tone for the discussion, Julius Richmond said that the IOM could "no longer offer promissory notes; it must begin to influence public policy more substantially." In support of this proposition, Richmond said that the IOM should begin to attack major problems such as national health insurance and biomedical research policy. Although Richmond did not know it, his priorities were exactly those of the Board on Medicine, which had singled out the same two problems for attention. Listening to the discussion, Hamburg floated an idea that he had been considering for a long time. He would ask the staff to "map out the terrain" in five major policy areas, giving staff a definite role in the "new IOM" and staking out the major issues that should be addressed. Hamburg thought that the five areas should be (1) health services, with special attention to national health insurance; (2) health sciences policy; (3) prevention of disease; (4) education for the health professions; and (5) mental health. This proved to be the major idea that emerged from the meeting.[17]

In a communication from Hamburg to IOM members in September 1976, he synthesized the staff, committee, and Council discussions of the IOM's mission. Here again, he announced his intention "to map the terrain of health by means of the multiple perspectives so distinctively available in IOM." In the past, IOM studies had been initiated by Congress and the executive branch, but the terrain maps would make it easier for the IOM to initiate its own studies. Although Hamburg did not believe that the IOM's five principal benefactors would continue to provide the sort of flexible money that would make such initiatives possible, the IOM might be able to obtain foundation grants in "particular program areas."[18]

Reorganization

The upshot of the long discussions was a reorganization. In March 1977, the Institute of Medicine created six operating divisions, each with its own staff director and its own advisory board of IOM members and other experts. The divisions corresponded to the categories in which the Program Committee had considered proposals and the subjects of the terrain maps. Before the reorganization, staff had worked on projects of the moment without a permanent assignment. The new plan made it possible for staff members to develop specialties and for the entire IOM program to have more coherence from one year to the next.

The titles of the divisions and their areas of responsibility underwent constant change as the health care policymaking environment, available staff, and available funds changed. In the original six-division scheme, the Division of Health Care Services focused on topics related to health services research, such as the financing of health care and issues of health care quality. The Division of Health Manpower and Resources Development concentrated on such issues as education and the proper distribution of medical specialists. Elena Nightingale, who was both a Ph.D. and the only M.D. among divisional directors, headed the Division of Health Promotion and Disease Prevention, which examined the impact of biological and social factors on disease and disability. The Division of Health Sciences Research was responsible for projects that related to "the conduct and support of biomedical and behavioral research." The Division of International Health worked on projects that related to health and economic development, and the Legal, Ethical, and Educational Aspects of Health Division was a catchall category to cover everything else.[19]

The creation of program divisions and the initiation of terrain maps transformed the way in which the Program Committee did business. What before had been an unstructured discussion of disparate projects became a much more disciplined conference on the six program areas. As Cecil Sheps, a public health specialist from the University of North Carolina, put it, every proposal had some value, but the "maps" would guide the committee in choosing among the proposals. David Hamburg himself chaired the group in charge of writing the terrain maps in health science policy and in prevention. Karl Yordy coordinated the committee working on health services policy. Drafts of these documents sparked spirited discussion within the Program Committee on an appropriate IOM role. In the area of health sciences policy, for example, the IOM might assess public participation in the field and ways in which the distribution of research dollars hampered interdisciplinary research. In the area of health promotion and disease prevention, the IOM might study the cultural factors that provided incentives for people to behave in healthful ways.[20]

By 1979, Dr. Theodore Cooper, the head of the Program Committee, could envision an entirely new role for this committee. No longer would it focus on specific studies. Instead, it would take a long-range view of health developments in the United States and identify "broad policy issues of high significance." Meanwhile, the division advisory boards would focus on specific projects that should be addressed in a particular year.[21]

By the time Cooper spoke, the division structure had already been altered. At the end of 1978, David Hamburg announced that the catchall division for medical ethics and other concerns would be abolished. In its place would come a new Division of Mental Health and Behavioral Medicine, an appropriate enough choice given the interests of the current administration and of David Hamburg himself. Another change came in IOM's management structure. As before, the president would be assisted by an executive officer. Since the departure of Roger Bulger, Karl Yordy had assumed this role. Just below Yordy in the organizational structure would come two program operations officers: Stan Jones, who had once worked for Senator Edward Kennedy (D-Mass.), would focus on program development and serve as principal staff to the Program Committee; Richard Seggel would focus on program operations and serve as principal staff to the Finance Committee. The division directors, in turn, would function as staff to the advisory boards in their particular areas.[22]

The new organizational structure achieved at least three separate goals. First, it increased the participation of members in IOM activities. Not only would members be able to serve on the Program Committee, they also would be able to serve on or make suggestions for an advisory board in their particular specialty area. Second, the new structure simultaneously expanded the role of the staff, because staff would have to guide the advisory boards and help in the creation of terrain maps. Finally, it enabled outside funders and other interested parties to observe what areas of inquiry the IOM regarded as most important and hence what projects most deserved to be funded.

Clearing the Pipeline

Before the new system could go into full effect, the IOM had to absorb the old studies that were in the pipeline. Perhaps the most important of these concerned the formulation of a manpower policy for primary care, a subject that would become a recurrent motif in the Institute's program. Begun in 1975 with funds from both the Robert Wood Johnson and the Kellogg Foundations and released in 1978, the study experienced many of the tensions that were evident in the IOM during this period. It fell somewhere in the middle on the scale between a large data-gathering study and a policy pronouncement. About a year into the project, Harvey Estes, the head of Duke's Department of Community and Family Medicine, who chaired the steering committee, asked the Council to increase the funds for the project so that the committee could develop estimates of the manpower required to supply an appropriate amount of primary care. The Council, wary of large data-gathering efforts, balked at the request and instructed the steering committee to reach conclusions based on the available evidence. When the completed study came before the Council for review, some Council members questioned whether the recommendations were internally consistent. They ordered the steering committee to change the order in which it presented its recommendations so as to convey "continuity of thought."[23]

The final report took as its objective the development of an "integrated" primary care manpower policy, although the steering committee admitted that it had undertaken no original research. Instead, the committee relied on staff papers, an open meeting at which interested organizations presented their positions, and steering committee members' "own expertise." The basic approach was to develop a working definition of primary care—"accessible, comprehensive, coordinated continual care delivered by an accountable provider of health services"—and then suggest ways in which to create an adequate supply of such care. The committee stressed that the answer was not to increase the number of medical school students but rather to alter the incentives for physicians and other health care practitioners to enter primary care fields. Payment differentials between primary care and other services should be reduced. The national goal for the percentage of first-year residents in primary care fields should be increased. All medical schools should direct, or have an affiliation with, at least one primary care residency program. When medical schools admitted students, they should favor those who wanted to go into primary care fields, and all medical schools should provide undergraduates with clinical experience in a primary care setting as well as training in epidemiology and other aspects of the behavioral and social sciences that were relevant to patient care. What distinguished the report from a laundry list of recommendations was that it included a checklist of the steps that medical schools could take to implement the changes advocated by the committee.[24]

Aware of the need to distribute the results of its studies, the IOM made efforts to place a summary of the study in a good journal. The article appeared in the *New England Journal of Medicine,* which is read by doctors in many different specialties, soon after the formal release of the report. It carried the names of three authors. The first two were members of the project staff and the third was Harvey Estes. In this way, the IOM not only disseminated the results of its study but also provided an outlet for the creative energy and professional advancement of the staff.[25]

Another IOM study of this period stemmed from a congressional request to investigate quality assurance programs. The programs reflected Congress's desire to make sure that Medicare and Medicaid patients received appropriate services from health care providers, as well as an interest in using the Medicare and Medicaid programs to contain health care costs. In 1972, Congress authorized the establishment of professional standards review organizations, and in 1973, it mandated a study of alternative mechanisms for health care quality assurance. Long negotiations between the IOM and the Department of Health, Education, and Welfare ensued, with the result that the study did not appear until 1976. The steering committee, under the direction of Dr. Robert Haggerty, a professor of health services and child health at Harvard, concluded that there simply was not enough available information to demonstrate the effectiveness of quality assurance programs. The committee offered suggestions, such as targeting reviews on "questionable patterns of care," to increase the effectiveness of existing review programs. As part of the same project, the IOM also studied the reliability of information on abstracts of hospital discharge records, finding wide discrepancies between the results obtained by an IOM team of researchers and the results in the hospital discharge records.

This study focused on the inner mechanics of health policy and the technical aspects of health care and answered questions posed by Congress, rather than by IOM members themselves. Within a very limited circle, such as the National Standards Review Council and professional conferences on disease classification, the study received wide play. The study fit the traditional NAS and National Research Council (NRC) model in which the Academy advised the government on technical and scientific issues. Indeed, Philip Handler welcomed competition between the NRC and the IOM in responding to requests to the Academy from the government. Even as Hamburg presided over the dissemination of the study on quality assurance, however, he hoped that the IOM would be able to break free of this mold and initiate its own studies.[26]

The Malpractice Study and the Polio Study

The medical malpractice project exemplified the sort of study that the IOM had wished to do for a long time. First discussed in the era of John Hogness, it came to fruition during the presidency of David Hamburg. Like many IOM studies, it took a great deal of time to complete and involved a considerable amount of turmoil within the staff and the steering committee. The final study, for all of the effort that went into it, turned out to be a brief report with few policy recommendations.

The project began when John Hogness appointed an Ad Hoc Committee on Medical Malpractice, which met for a few months in 1973, at the time the HEW Secretary's Commission on Medical Malpractice was about to issue its report. This commission recommended that a uniform body of legal rules governing medical malpractice be created for courts to use throughout the country. The IOM Ad Hoc Committee found many deficiencies in the report, such as its failure to address the deterioration of the physician–patient relationship and the creation of conditions in which patients might well bring a tort action against their doctors. The committee asked that there be an IOM policy statement on this subject. The project then languished for lack of funds. When David Hamburg arrived in

1975, he found the project moribund but the subject compelling and asked IOM staff to find out what had happened in the field of medical malpractice between 1973 and 1975. Staff members Barbara Cohen, who had both Capitol Hill and White House connections, and Michael Pollard, who had both a law and a public health degree, made a round of Washington visits and came up with a new proposal for a study in the area of medical injury compensation. The Henry J. Kaiser and the William and Flora Hewlett Foundations contributed the money to fund the project.

John Hogness agreed to chair the steering committee, which met for the first time in August 1976, in part because he wanted to show his continuing interest in IOM affairs and in part because the subject was one about which he knew little and wanted to learn more. The committee contained an interesting mix of medical practitioners, such as Jeremiah Barondess, a professor of medicine at Cornell University Medical College; social scientists, such as Stanford sociologist Richard Scott and Wisconsin economist Burton Weisbrod; and legal scholars, such as Guido Calabresi of the Yale Law School (who later became its dean). The group also included Jonathan Spivak, the *Wall Street Journal* correspondent who followed events in the health care field closely.

Like nearly all IOM studies of this era, the work of the Medical Malpractice Committee became delayed. Barbara Cohen, the staff director, left the IOM, and Michael Pollard stepped in to take her place on the study. The budget for the project was tight, and Hamburg had to ask the sponsors for another $30,000 to complete it. Guido Calabresi received the staff draft of the final report in July 1977 and objected so strongly to it that he decided to resign from the steering committee. He complained about the report's loose grasp of economic theory, as in the statement that the cost of rising insurance premiums was passed along to customers.[27]

The final report, dated March 1978, stated at the end of 64 closely reasoned and heavily annotated pages, that "the focus of public policy and research that relate to medical injury should not be unduly restricted to instances of medical malpractice." Instead, policymakers should concentrate on "the incidence of medically related injury, possible techniques for preventing injury, and financially sound methods for compensating injured patients." In other words, the emphasis should not be on fault, as it was in the present tort system, but rather on prevention and on outcomes. The system should strive to prevent disability and alleviate its consequences, not encourage the deterioration of the physician–patient relationship. Beyond these important insights, the report offered few specific suggestions for preventing or compensating medical injuries. The report was best read as a primer on compensation for medical injuries, not as a statement of policy or a research report.[28]

The medical injury report cast the IOM in the role of student rather than teacher. When the organization assumed the part of an expert arbitrating a dispute, it obtained better results. The polio vaccine study, another of the studies completed before David Hamburg's new organizational scheme took hold, illustrated how the IOM could use its expertise to resolve public policy disputes.

Polio vaccinations were one of America's greatest public health triumphs. After the introduction of the Salk vaccine in 1954, new cases of the disease virtually disappeared. By the 1970s, the number of new cases each year could be counted in single digits. Public health officials worried as much about people not getting vaccinated as they did about the safety of the vaccine itself. For greater ease of administration, authorities substituted the use of an oral vaccine for the injected vaccine in 1962. All a person had to do to prevent polio was swallow a sugar cube. Unlike the Salk vaccine, the Sabin oral vaccine was prepared with what scientists called "attenuated live virus." That meant that although the oral vaccine was very safe, it did lead to very occasional cases of polio, perhaps 44 cases between 1969 and 1976, among people who took it or came into contact with someone who had taken it.

The problem acquired more visibility in 1976 in the wake of the effort to prevent an epidemic of swine flu. A number of people who were vaccinated against the flu suffered adverse reactions, creating a major public scandal and raising questions about whether the government or the vaccine manufacturer was liable for damages. In September 1976, an HEW official testified before a Senate subcommittee that the government was having trouble entering into contracts with private companies for the manufacture of vaccines of all types, including those against measles, rubella, and polio. A public health crisis loomed, as officials worried about a shortage of vaccine.

Against this background, Dr. Theodore Cooper, the Ford administration's Assistant Secretary of Health (and later, head of the IOM Program Committee), approached David Hamburg about responding to a request from Senators Edward Kennedy and Jacob Javits (R-N.Y.): Would the IOM look into the relative merits of live (Sabin) versus killed (Salk) polio vaccines? Because of the short congressional deadlines, the job would have to be done quickly; Senator Kennedy wanted the report in the spring of 1977.[29]

The IOM responded with alacrity. In less than two months, Elena Nightingale of the IOM staff made arrangements for two committee meetings and a two-day workshop. She and Hamburg persuaded Bernard Greenberg, dean of the School of Medicine at the University of North Carolina at Chapel Hill, to chair the study and recruited a steering committee that contained Fred Robbins, a Nobel laureate for his research on polio and a future IOM president, and Byron Waksman, a professor of pathology at Yale. The time line for the report was exceedingly tight. The committee would meet for the first time on February 10, 1977, and deliver its report on April 15.[30]

The committee met its deadlines. It convened in February to develop a working plan and to organize an international workshop that would take place in March. The workshop gave the committee the chance to consult with authorities from countries such as the Netherlands, which had not switched from the dead to the live vaccine. It also provided the committee with an opportunity to assemble working groups on practical questions, such as how to be obtain informed consent from those who received polio shots, and on the safety and efficacy of polio vaccines in the United States. After these groups departed, the committee assembled in executive session and reached decisions on the major public policy questions. On April 6, 1977, the committee presented its findings at a scientific conference devoted to immunization; nine days later, exactly on schedule, it delivered its report to the Department of Health, Education, and Welfare.

The committee recommended that the United States continue the use of the Sabin oral vaccine as its principal means of preventing polio. It added a number of caveats. Those with heightened susceptibility to infection or adults who were being vaccinated for the first time should continue to take polio shots. The committee also advised that there be a new round of immunizations, to be given orally to all children as they entered seventh grade. The committee hoped that this practice would protect the children in their later years when they became parents and, in this manner, eliminate cases in which parents contracted polio from their immunized children. The committee noted that the country could achieve a higher rate of immunization against polio. As of 1974, for example, only 45 percent of all nonwhite children had been vaccinated against polio. The committee hoped that the nation could reach a 90 percent immunization rate within a few years. To help obtain this goal, the committee advised that all liability from the immunizations, except in cases of gross negligence, be assumed by the government.[31]

The polio vaccination study attracted a great deal of attention. The *New England Journal of Medicine* ran an article that summarized the study, as did the widely circulated *Scientific American*. Elena

Nightingale reported that she had been besieged by requests for copies of the study from groups around the world, including the Belgian government and the World Health Organization. Perhaps the greatest achievement, according to Nightingale, was that the study helped to allay the hysteria that had followed the swine flu vaccine debacle. The study, noted Nightingale, "has aroused quite of bit of positive interest in poliomyelitis vaccination" and, she might have added, in the Institute of Medicine.[32]

The Embrace of the Carter Administration

The computed tomography scanning project resembled the polio vaccine study. It was to be a "rapid, yet incisive appraisal of the proliferation of computerized tomography scanners in medical practice." The Blue Cross Association, which served as the umbrella organization for the many local Blue Cross hospital insurance plans across the country, sponsored the study because of its desire, as study director Judith Wagner delicately put it, for the "judicious use of technological innovations in medicine."[33] In other words, Blue Cross hoped to put appropriate limits on an expensive procedure. The IOM study helped establish these limits.

The subject of cost containment also figured prominently in a statement about controlling the supply of hospital beds, which the IOM released just before the 1976 presidential election. As with other IOM studies, a long time was needed to acquire the necessary funding and to find an appropriate focus for the project. The completed report stated unequivocally that "significant surpluses of short-term general hospital beds exist or are developing in many areas of the United States and that these are contributing significantly to rising hospital care costs." The study panel, headed by Robert Heyssel, director of the Johns Hopkins Hospital, recommended that within five years there be a reduction of at least 10 percent in the ratio of short-term hospital beds to the population. The statement provoked a dissent from Harold Cross, a general practitioner from Hampton Hills, Maine. He argued that "to cap the process at some arbitrary bed limit before assuring minimal care is irresponsible and gives the appearance of having provided a solution."[34]

Among those who very much wanted to see a solution to the problem of health care cost containment was Jimmy Carter. The President realized that any campaign to create national health insurance would have to be accompanied with measures to reduce the rate of growth of health expenditures. It was for this reason that the article in the *Washington Post* about the CT study had caught his attention.

President Carter and HEW Secretary Joseph Califano already had the full attention of the Institute of Medicine. David Hamburg and the IOM staff greeted the return of a Democratic administration with enthusiasm. Although the organization tried hard not to favor one party over another, it nonetheless had more in common with the Democrats than the Republicans. The Board on Medicine had been created in the shadow of the Great Society and in the expectation that the social experimentation of the 1960s would continue into the 1970s. Richard Nixon's election stunted the hopes of many Board on Medicine members and staff for such things as federally administered national health insurance.

In the Nixon and Ford eras, IOM staff eventually developed and maintained many close ties to administration officials, such as HEW Secretary Elliott Richardson and Assistant HEW Secretary Merlin DuVal. During the Ford administration, in particular, David Hamburg worked closely with Theodore Cooper, the Assistant Secretary of Health, and Guy Stever, the President's science adviser. After Cooper was fired by Califano, he received an invitation from Hamburg to come to the IOM as a visiting scholar. As these examples illustrated, party affiliation almost seemed not to matter in the health care field: professional identity often superseded partisan loyalty. In this spirit, the Ford administration named Donald Fredrickson director of the National Institutes of Health (NIH), and a similar bipartisan spirit prevailed in Congress. Nonetheless, IOM officials realized that a Democratic administration offered more chances for collaboration than did a Republican. David Hamburg said as much in March 1977, when he told the IOM Council that he and his staff had been approached for advice by the new administration. He thought that such contacts might give the IOM more chances to become involved in the formulation of health policy.[35]

The embrace of the Carter administration sent the IOM off in two different directions. On the one hand, the IOM wanted to be of service to an administration that appreciated its advice. On the other, the IOM wanted to use its divisional structure, which had just been put in place, to generate its own agenda. Just as the Hamburg reorganization took hold, the arrival of the Carter administration upset its workings.

At the time, few people saw any problems in the IOM maintaining a close relationship with the Carter administration. The opportunity to be on the inside track only increased the chance for the IOM to play an important role in health policy at a time when this policy appeared to be at a point of momentous change. According to this view, the IOM would maintain its integrity even as it stepped closer to the center of power. As Hamburg had stated in his final president's report to the IOM, "We must take a sympathetic interest in government efforts and try to be helpful where we can. But the Institute can be most helpful in the long run if our actions have the degree of insight and objectivity." So the IOM would continue to speak truth to power, only now power would listen.

In IOM Council meetings, David Hamburg reported on the latest developments in the Carter administration and encouraged IOM participation in them. He also invited administration officials to give presentations at Council meetings, a practice that Hogness had followed but that had lapsed in more recent years. In March 1977, for example, Hamburg told the Council that an ad hoc group had been convened to offer suggestions to the Presidential Commission on Mental Health. Even though the commission had not yet been formally appointed, Hamburg—whom one staff member described as "psychiatrist to Washington, D.C."—to whom everyone, even cabinet members, turned for inspiration, already knew that the IOM would be asked to prepare a study for the commission. Hamburg also reported on a bipartisan congressional request that the IOM review promising leads in the field of international health. Senator Edward Kennedy and his congressional colleagues from both parties expressed an interest in how America's "medical research capacity" and its "experiences in organization and delivery of primary and preventive care" might benefit other nations, an interest in the international dimensions of health that the Carter administration shared. In May, members of the IOM Council had dinner with Secretary Califano and met with Donald Kennedy, commissioner of the Food and Drug Administration (FDA), and Hamburg's former colleague at Stanford. In the same month the Council learned that Julius Richmond, who had served as an interim head of the IOM between Fredrickson and Hamburg, was the Carter nominee for the position of Assistant Secretary of Health.[36]

Working for the Carter Administration on International and Mental Health

Each of these administration ties led to work for the IOM. Already interested in the field of international health after his experiences in Africa, Hamburg urged the IOM Council to accept the congressional invitation to do research on this subject. At its own expense, the IOM

convened an International Health Committee that prepared a report
on research opportunities in the field of international health. The field
became integral to the IOM's basic activities, with a program division
devoted to it. In 1978, for example, Hamburg made international
health a major focus of the IOM's annual meeting. Although he was
not sure how much interest members had in the topic, he believed
that "gravity of disease conditions in other parts of the world" merited
"serious attention from the Institute."[37]

In July 1977, the IOM Council learned how important
international health activities were to President Carter. It heard from
Peter Bourne, a Carter White House staffer and former student of
David Hamburg's, that health was a "pivotal concern in the
administration's strategy for improving relations with other
countries." Bourne said that in the President's human rights policy,
"problems in food supply, health care, and shelter were likely to
receive as much emphasis in dealings with other countries as civil
rights."[38] Congress mirrored the President's interest. Aware that the
IOM was doing work in the field, members of the Senate Committee
on Human Resources attached an amendment to a public health bill
that provided for an IOM study "to determine opportunities, if any,
for broadened Federal program activities in areas of international
health."[39] This study became the first formal product of the IOM
Division of International Health.

The study itself appeared in April 1978. With the steering
committee chaired by John Bryant, director of Columbia's School of
Public Health, the report reflected the work of four subcommittees
that, taken together, provided a panoramic view of the field. One
group looked at the major diseases of low-income countries. A second
investigated the ecological, socioeconomic, and cultural factors
involved in health. A third studied environmental control programs
and health education possibilities, and a fourth attempted to
ascertain the feasibility of U.S. involvement in international
programs to meet the problems identified. The resulting report
identified the "major policy and organizational problems and
constraints which currently hamper the U.S. government's international health activities." The committee concluded that "the current
base of knowledge and experience provides the possibility of
ameliorating many [health] problems by commitments of realistic
amounts of resources by both developing countries and economically
advanced countries." If some of the rhetoric was self-serving, as in a
recommendation that the Agency for International Development
(AID) enter into relationships with U.S. academic institutions to
undertake research and development activities in the field of

international health, the report nonetheless marked a credible IOM entry into a new field of endeavor. This case was one in which the interests of the Carter administration combined with those of David Hamburg to shape the IOM's development.[40]

Something similar happened in the area of mental health. Once again, the interests of the Carter administration coincided with those of David Hamburg to produce a series of IOM studies. These concentrated on the links between health and behavior, which had long been a focus of Hamburg's research. Before he left Stanford to take the IOM job, Hamburg told his colleagues that one of his top priorities was to try to get the IOM "to look at behavioral aspects of health over the whole range of health." The work done by IOM for the President's Commission on Mental Illness presented Hamburg with an opening to pursue this interest. He used this work as a base from which to negotiate with Carter administration officials for an expanded IOM role in the field of behavioral medicine. As Hamburg described the process, the government and the IOM mixed and matched interests. The IOM had things that it viewed as "central to our agenda and the government may buy some of that, at least in modified form. On the other hand, they may come at us with questions that are salient to them and that don't particularly fit our agenda and yet, within the mandate of this institute, we pretty much have to take up because they're not farfetched, they're not disreputable, they're difficult and so on. So you end up with that kind of funny mosaic of our agenda and theirs." Hamburg, Julius Richmond, Donald Fredrickson, and the heads of the various NIH and Alcohol and Drug Abuse and Mental Health Administration (ADAMHA) institutes created the mosaic that became the IOM project "Health and Behavior: A Research Agenda."[41]

The project consisted of a series of IOM conferences, each of which generated its own report, on specific questions in the field of health and behavior, followed by a volume that synthesized the conference results, suggested promising research leads, and integrated "available information into a perspective of the frontiers of the biobehavioral studies." As one indication of David Hamburg's personal interest in this project, he decided to chair the steering committee, which met for the first time in November 1979. Other committee members were leading figures in the fields of psychology, psychiatry, and psychobiology. Because so many different federal agencies were involved in the task, the contract for the project was not signed until after the project had begun. The final report did not appear until the summer of 1982.

The first conference volume, on the links between smoking and behavior, followed an interest shared by Surgeon General Julius Richmond and HEW Secretary Joseph Califano. It featured an introduction by Richmond in which he noted that the volume supplemented the work he had undertaken for a 1979 report on smoking and health. Both the conference volume and Richmond's report emphasized the disparity between the large amount of biological research that showed the deleterious effects of smoking and the small amount of behavioral research on what caused people to smoke and what might encourage them to stop.[42]

Five more conferences, each on a topic of interest to one or more federal agencies, followed the conference on smoking. Robert J. Haggerty, president of the William T. Grant Foundation, chaired the Conference on Combining Psychosocial and Drug Therapy. The idea emerged that "behavioral science has a powerful role to play in conjunction with traditional biologic research at the levels of both the individual patient and of society if we are to improve the nation's health."[43] In a subsequent conference, participants tried to bring the concept of "social disadvantage" into the mix. They noted, for example, that rates of severe mental illness were higher among members of the lower social classes and argued, in the manner of social scientists from the 1960s, that "inadequate resources, low status jobs, social stigma and inadequate education interact with differential immunity, nutrition, environmental risks and coping styles to create a 'circle of disadvantage'."[44] As the social scientists had learned earlier, this was tricky terrain, invoking memories of Daniel Moynihan's "tangle of pathology" in the black ghetto. Indeed, the boundary between the personal and the social realms was difficult for participants in all of the conferences to define. In the field of aging, for example, researchers had to distinguish among events that were biologically determined, culturally determined, and personally determined.[45] In a similar vein, Leon Eisenberg, summarizing the results of the Conference on Infants at Risk for Developmental Dysfunction, said that efforts "to understand . . . the various risk factors experienced by both mothers and children during pregnancy . . . and early infancy make sharply evident the need for research that integrates sociobehavoral with biomedical paradigms."[46]

The conferences yielded a plethora of information that made for a fascinating final report. Within the Academy complex, furthermore, only the IOM could produce such a report, because of its ability to engage in interdisciplinary work and combine biology with behavior. The organization of the National Research Council, by way of contrast, resembled the traditional departments in a university, with all the attendant problems of producing interdisciplinary work. The IOM study found that as much as half of the mortality from the 10 leading causes of death in the United States could be traced to a person's life-style. In the typical hospital population, one encountered a disproportionate number of people who had engaged in alcohol abuse, cigarette smoking, or overeating to obesity. Pregnant mothers spread the problems to the next generation. Cigarette smoking, for example, doubled the risk of having a low-birthweight infant; mothers who drank heavily faced a far greater chance that their babies would suffer from fetal alcohol syndrome. All in all, "relationships among the stress of life events, social supports, and various styles of coping" offered a "rich area of research opportunity." Such research would be informed by the knowledge that "for many chronic diseases, drugs alone are not enough." Through this research, health care practitioners would come to understand just how "social and psychological influences affect the disease course and prognosis." The project amply demonstrated that "the leading causes of illness and death have substantial behavioral components, so approaches to preventing or managing them must include a strong biobehavioral perspective."[47]

Specifying the exact links between behavior and disease presented many difficulties, as the IOM discovered when it responded to a request from the Office of Science and Technology (OST) in the White House that it study the relationship between stress and behavior. White House staff members had identified stress as, in the jargon of the day, a cross-cutting issue that related both to mental health and to the subject of illness prevention. The White House brought together the National Science Foundation (NSF), the National Institute of Mental Health, and the National Institute of Aging to fund the study, which ran concurrently with the one on health and behavior. An elaborate undertaking, the stress study involved the use of many panels and consultants; as many as 100 scientists participated.[48]

On November 26, 1979, when David Hamburg greeted the members of the steering committee, headed by Carl Eisdorfer, president of Montefiore Hospital and professor of psychiatry and neurosciences, he told them that they faced an arduous process. They had to do more than simply denounce stress as a contributor to mortality and morbidity; instead, they had to try to uncover some of the linkages between stress and ill health and to report on those areas in which more research was required. He cautioned that interest in this subject was high. Indeed, he later wrote "no aspect of health and disease elicited more interests among leaders of the United States government" than the subject of stress. Hamburg was tempted to elevate stress into a metaphor for the Carter years, with their series of wrenching events such as the radiation accident at Three Mile Island and the Iran hostage crisis. The press would therefore seize on the report and be eager to find a simplistic theme with which to characterize it, but the IOM and NSF committees that reviewed the report would demand more. Describing this review process, Hamburg said that "there's a lot of feedback . . . and some of it is jarring. I mean it is quite difficult. . . . [T]o retain a high degree of consensus in the face of a stiff critique is not simple, but I think it is very important."[49]

As Hamburg had predicted, the report ran into difficulties in the review process that delayed its release until the end of 1981. Although three of the IOM reviewers approved the report, two had strong reservations. They believed that the report failed to support its recommendations and complained that it was full of "tautologies, nebulous notions, and old research." "All of these statements are true," conceded the IOM staff members in charge, "the committee was not in a position to remake the stress field, which has been characterized by just those problems for many years." What the report tried to do was present results that "transcended those limitations" and replace "tautologies with insights, old research with new and well-designed studies, and vague ideas with hard data." The final report suggested that investigators move "beyond questions about whether stress can affect health and explore more fully the mechanisms through which stressors might produce such consequences." They should do so even though the stress response was a "complex, interactive process."[50]

The process of collaborating with and advising the Carter administration extended well beyond the area of health and human behavior. In fiscal year 1979 alone, the Institute reviewed HEW's planning process, reported on food safety policy and on the health hazards of sleeping pills for the FDA, investigated the research agenda of the National Institute on Alcohol Abuse and Alcoholism, and studied health in Egypt for AID.[51] It also tried to be of service in the legislative battle over national health insurance.

Working with the Carter Administration:
The Dental Study and the Surgeon General's Report

A good example of such service concerned the study of dental care that the IOM started in 1976 before the Carter administration took office. From the beginning of this study, the IOM had national health insurance in mind. It would be part of the general effort to examine "selected issues in national health insurance," which was featured in the IOM's 1976 program plan. Considered from this angle, dental care represented a neglected area of study. Most national health insurance bills included dental care for children up to age 13 on the theory that providing access of this sort would pay substantial dividends in the form of improved health. The IOM hoped to put this theory to the test and make other suggestions for just how dental care might be included in a national health insurance bill. Most of the analysis for the study was to be done by Chester W. Douglass, a professor of dentistry at the University of North Carolina at Chapel Hill and a former Robert Wood Johnson congressional fellow at the Institute of Medicine. The Kellogg Foundation agreed to support an 18-month study.[52]

Delays plagued the study from the start. On April 20, 1977, Hamburg invited Julius Richmond to chair the steering committee. When Richmond was nominated to be Assistant Secretary of Health, he had to resign and Dorothy Rice took over. As the project proceeded, the Carter administration began to formulate and Congress to consider a national health insurance bill. On May 25, 1978, Stan Jones, the IOM's program development officer and former Hill staffer, reported that President Carter would issue a statement of principles on health insurance in June. A benefit package would be specified in the fall. Jones advised the committee to release an interim report during the summer in an effort to influence the debate.[53]

The committee and its staff duly prepared such a report with very specific recommendations: for example, that national health insurance should provide individual preventive services for everyone, including routine prophylaxis. This report met with such strong disapproval from the IOM Council that it was never issued. The IOM Council reviewed the report as though it were a rigorous scientific study and wanted to see more evidence on such questions as whether a dental checkup on a regular basis would prevent health problems, fearing that, without such evidence, the report amounted to little more than a list of recommendations. Others questioned the economic logic of the report. Walter McNerney of the Blue Cross Association said, for example, that his experience showed that the removal of economic barriers contributed significantly to increases in utilization by the covered population. Hence he believed that the cost estimate of $1 billion for the coverage was unrealistically low. Faced with this criticism, the group decided to abandon the interim report and move on to the final report.[54] Disagreements between Chester Douglass and the steering committee led to more delays. Money for the project ran out and the IOM had to ask for more. The IOM brought in William Fullerton, a retired Social Security policy analyst and congressional staff member, and Stan Jones, who had left the IOM to become a private consultant, to act as "script doctors" and rewrite sections of the report. The final report did not appear until the end of 1980, long after the window for passage of national health insurance had closed.[55]

It was clear that the rhythms of the IOM did not always mesh with those of Congress or others close to the political action in Washington. Part of the problem was that the IOM sought to be both an outside advocate and an inside collaborator in its relations with the Carter administration. In at least one instance, the IOM offered, in effect, to staff a major Carter administration effort. Differences in outlook between the IOM and HEW made for a strained collaboration.

The project stemmed from Surgeon General and Assistant Secretary of Health Julius Richmond's desire to issue a report on prevention as a major theme of health policy. In February 1978, the IOM held a conference on health promotion and disease prevention, a subject that had received a great deal of attention from the Carter administration. In May 1978, Richmond asked the IOM to prepare a report, based in part on the conference, that summarized the "state of the art." This led to a formal proposal from the IOM in July and a formal contract from HEW by the middle of August. In the meantime, Secretary Califano announced that Richmond would issue a major report in the autumn "to help Americans fight against obesity, alcoholism, and many other costly everyday health problems." The IOM staff, assisted by a special advisory committee, rushed to get something to Richmond by October. The IOM also commissioned a series of papers on particular aspects of the subject, such as reducing tooth decay in children or lowering the number of motor vehicle accidents. In this way, the IOM hoped to be "as helpful as possible to the Surgeon General of the United States in writing a Surgeon General's Report on Prevention."[56]

As more units of the Department of Health, Education, and Welfare became involved in the project, the Surgeon General's report on health promotion and disease prevention became delayed. Serious disagreements arose between HEW and the IOM about how to handle the report. The IOM objected in particular to the way in which the surgeon general's staff "emphasized individual responsibility for health way out of balance with governmental and social responsibilities." The IOM contemplated withdrawing from the project and publishing the report it had prepared and the studies it had commissioned on its own. In the end, David Hamburg and Julius

Richmond managed to restore order. *Healthy People: The Surgeon General's Report on Health Promotion and Disease Prevention* appeared in August 1979, and Joseph Califano signed it as one of his last official acts as Secretary of HEW. As Hamburg explained, the book drew heavily on the material that IOM had prepared but that, "naturally, DHEW has put its stamp on the material." A second volume of the Surgeon General's report consisted entirely of the background papers that the IOM had commissioned.[57]

During the summer of 1979, Jimmy Carter delivered his famous homily on the nation's wavering sense of purpose that the press dubbed the "malaise" speech. In the wake of this speech, the President announced a shakeup of his cabinet, which led to the dismissal of Joseph Califano. According to a predetermined schedule, the Surgeon General's report was released just after Califano had been fired. As a result, it received little of the fanfare that Califano might have brought to it. The bold call for a "second public health revolution" to prevent all disease tended to go unheeded.

In fact, the Surgeon General's report revealed an important fault line in the nation's political ideology. In the past, Democratic administrations had emphasized measures that the government could take to improve the nation's health, such as the creation of Medicare or increased investment in medical education and research. The Carter administration inherited many of these projects from past administrations, in particular the desire to shore up the financing of health care through national health insurance. Unlike the other administrations, Carter and Califano started with the primary goal of cost containment as a necessary precondition for guaranteeing universal access. Prevention was a key to cost containment, and Carter administration officials tended to see prevention as an individual, rather than a governmental, responsibility. The Surgeon General's report seemed to confirm the fact that, as the *Washington Post* reported, "people can do far more to improve their health by acting themselves than they can by waiting for symptoms and then going to doctors."[58] The IOM committee in charge of advising the surgeon general was not prepared to push the prevention line this far. Committee members continued to see an important role for government in disease prevention and health promotion.

The dental study and the Surgeon General's report on prevention revealed that, however much the IOM wanted to be of service, it was removed from politics. It could get only so close to a particular administration before the differences between its style of operation and those of a more politicized organization became apparent. In the case of the Carter administration, the departure of Califano, combined with the demise of many of the items on the administration's agenda, led to an increased separation between it and the IOM. Hamburg hoped that the IOM could be as helpful to Patricia Roberts Harris, Califano's replacement, as it had been to Califano. As David Hamburg later noted, Califano early saw the IOM's value and sought to work with it. As Califano came to see the increasing importance of health as an area of social policy, he developed an attachment to the IOM that continued after he left office. Califano became an IOM member and served on the IOM Council during the 1990s. Patricia Harris, for her part, held many fruitful discussions with Hamburg and proved to be very supportive of the IOM's international health efforts. Still, her time in office was brief and the possibilities of collaboration were fewer during her tenure than during Califano's.

Riddles of Power

Being close to power was exhilarating but not without its problems. To preserve its reputation, the IOM always had to fend off blatantly political requests. A good example in the Carter era concerned a request from Senator Daniel Inouye (D-Hawaii) that the IOM study the health effects of tourism. Staff members from the senator's Subcommittee on Tourism of the Senate Commerce Committee visited IOM staff member Karl Yordy and explained that they wished to explore ways in which the effective use of leisure might lead to improvements in health. If a link between leisure and good health were established, the government might mandate vacations in the same way that the Fair Labor Standards Act limited the standard working week to 40 hours. Yordy listened as politely as he could and talked of the many difficulties that an IOM study might pose, such as the problem of conceptualizing leisure. Undeterred, Senator Inouye sent David Hamburg a formal letter, noting that hearings in his subcommittee "raised the possibility of linkages between stress . . . one's ability to relax and various physiological and psychological symptoms." He even suggested that the government might want to promote four-day work weeks and mandatory paid vacations as part of its public health responsibility. Calling the IOM's academic and professional expertise "second to none," Inouye asked the IOM to help the subcommittee develop policy recommendations.[59]

The self-interest in this request was obvious. Inouye represented a state whose economic well-being depended on the health of its tourism industry. What he wanted, in effect, was to use the federal government to subsidize the economy of Hawaii on the recommendation of the IOM. The press had already begun to investigate this matter. Yordy told Hamburg that *Newsweek* had almost put the study in its "golden fleece" category, certifying it as a boondoggle. Wary of offending an important senator, Hamburg bowed out as gracefully as he could. He thanked Inouye for his letter and said that he personally wanted the IOM to do the study but it would require review by the Program Committee. In this way, he let the matter drop.[60]

Even on a project that the IOM and government officials both wanted, complications related to politics sometimes developed. A good example concerned an IOM conference on the subject of the care of terminally ill patients. At first, the National Cancer Institute (NCI) expressed an interest in funding such a conference. In a few months, however, it became apparent that the NCI would not do anything to help the project; IOM officials suspected that priorities within the NCI had changed. Then Joseph Califano gave a speech in which he embraced the hospice concept and announced plans for a national conference on the care of dying patients. National Cancer Institute officials immediately got in touch with the IOM and stopped work on the contract they had been negotiating. The IOM then began discussions with Califano's staff to see if it could play a role in his conference. As discussions progressed, IOM officials realized that HEW envisioned a grandiose conference, attended by more than 1,000 people, that would serve as a platform for Califano to follow up on his earlier speech. The IOM wanted no part of organizing such a conference and proposed instead that the IOM provide "credibility, objectivity, and intellectual input" by convening a committee to advise on the conference agenda and speakers. HEW rejected this idea, and the negotiations ended.[61] In this case, the IOM could not maneuver between its desire to study an issue and Califano's desire to publicize the case for hospice care.

Just after Joseph Califano was fired, an exasperated Karl Yordy wrote to his counterpart at the Kellogg Foundation that "obtaining HEW support during the last year has been such a struggle, in spite of great goodwill for the Institute at all of the policy levels within the Department, that it is difficult to imagine the situation being any worse."[62]

Fund-Raising

Karl Yordy and David Hamburg realized that government funds and government projects would never take the place of private

foundation   support.   During   Hamburg's   presidency,   the   core
foundations continued to finance the Institute, and the IOM managed
to add new patrons to the list.

When Donald Fredrickson announced his intention to leave the
IOM, a program officer at the Robert Wood Johnson (RWJ)
Foundation hastened to assure Karl Yordy that the RWJ staff
intended "to continue to support IOM regardless of the leadership
change." This proved to be the case. The Kellogg Foundation told the
IOM in 1975 that Kellogg would support only "specific project
activities," yet Kellogg continued its generous support, as a 1978
award of $375,000 to "help underwrite the Institute's annual program
plan" demonstrated. In 1978 the IOM also received a check for
$750,000 from the Andrew Mellon Foundation to "address policy
issues critical to the nation's use of finite resources for health care."
During the next year, a $105,000 check arrived from the Richard King
Mellon Foundation for the "leadership and initiative functions of
IOM."[63]

Even as David Hamburg hastened to reassure these foundations
that the IOM was still worth supporting, he also courted new ones. A
notable success came with the Charles H. Revson Foundation. Using
IOM Council member Lisbeth Bamberger Schorr as an intermediary,
Hamburg set up a meeting with Revson Foundation Director Eli
Evans. "I think our interests overlap in many ways," Hamburg wrote
to Evans after the meeting. Three months later, Hamburg learned
that the Revson board had appropriated $200,000 for the IOM to
conduct four Revson seminars on biomedical research.[64]

Although these donations helped, they still could not free the IOM
from debt. In July 1979, for example, the IOM Council learned that
the fiscal year would end with a budget deficit of $210,000. The
National Academy of Sciences offered to bail out the IOM but at the
cost of reducing its support for fiscal year 1980. This meant that the
IOM would have to cut back on its 1980 expenses.[65] A major part of
the problem stemmed from a ruling made by government auditors in
1978 that the IOM had to include overhead costs for private
foundation grants in its budget. In other words, if the IOM received
money from Robert Wood Johnson, part of this money would have to
go toward indirect costs or overhead. Not all of the money could be
spent on things such as salaries, meetings, or conferences; part of it
would have to go toward general upkeep of the Institute. This ruling
had the effect of reducing the amount of the foundation grants and
upsetting the IOM budget so that the IOM suddenly found itself with
a $750,000 debt. Although IOM staff members hoped that President
Handler would appeal the auditors' ruling, he preferred, in the

privately spoken words of one IOM staffer, "to chastise the IOM for its profligate and careless ways." Hamburg thought about compensating for the loss of income by obtaining what he described as "flexible kinds of support from federal quarters."[66] Such support proved very hard to obtain because federal agencies insisted on contracts for specific projects and the government did not make flexible grants in the manner of foundations.

Defining the Institute's Mission

The oversight that the National Academy of Sciences exercised over the Institute complicated any efforts to resolve the IOM's funding difficulties. Although relations between the NAS and the IOM tended to be calm in the Hamburg years, sharp differences of opinion did arise over such issues as the report review function. When John Hogness was IOM president, he developed a good working relationship with NAS Vice President George Kistiakowsky. As a result, the review of IOM reports by the Academy became routine. "We never had any difficulties," Hogness recalled.[67] This changed during the Hamburg years. In March of 1978, for example, Philip Handler sent Hamburg a long memo in which he pointed out his "vaguely negative" initial reaction to a report on medical technology, the disastrous history of the steering committee for the project, and the "undocumented assertions" that abounded in the report. He asked if Hamburg could regard the study as IOM's "best effort."[68] Hamburg hastened to make changes and thanked Handler for making the report "sounder."[69]

Such chastisement from Philip Handler rankled, as did the criticism of IOM reports by Saunders Mac Lane, who was a University of Chicago mathematician and NAS vice president, and chairman of the Academy's Report Review Committee. In 1980, Mac Lane told the IOM Council that the reports had run into difficulties because the IOM sometimes started "with the view of influencing policy rather than conducting a dispassionate assessment of fact." The resulting reports did not make explicit which of the findings stemmed from "informed judgment" and which rested on "hard evidence." At times also, conclusions reflected the opinion of the staff, not the results of investigations undertaken by the steering committee. In a similar spirit, Vincent P. Dole, vice chairman of the NAS Report Review Committee, cautioned the IOM not to become too involved in the political process when dealing with health policy problems.[70]

In response to this criticism, Hamburg pointed to the differences between basic and applied science. The social problems of the sort that the IOM addressed had a scientific content, "but all also entail uncertainty in the knowledge of factors that can influence outcome." If the IOM wished to tackle "the truly large, difficult issues of health and disease," as Walsh McDermott had hoped it would, it could not adopt the attitude of the pure scientist. Instead, the IOM had to consider the gray areas that lay beyond scientific certainty. As Fred Robbins, who succeeded Hamburg as IOM president, stated bluntly, "The Institute could not limit its study to the hard evidence and still carry out its mission."[71]

At the end of the Carter administration, it was still difficult to discern this mission. It was clear that the IOM walked a fine line between the scientific concerns of the National Academy of Sciences and the political preoccupations of policymakers. It also tried to sort out the most appropriate projects from those that its members suggested and those that the government brought to it. The fact that it could filter these proposals through separate divisions, each with its own advisory committee, helped lend coherence to the effort. Even so, Renee Fox, a University of Pennsylvania sociologist and member of the IOM Council, called the program plan for 1980 a "diffuse set of projects." Margaret Mahoney, head of the Commonwealth Fund and an IOM Council member who had done much to encourage philanthropic support for the Institute, said that "the time has come when [IOM] should think with some continuity so that people will associate certain kinds of activities with the Institute and no other organization." In this sense, the IOM, although it had discovered important themes such as primary care, health education, quality assurance, and prevention, had not yet found its niche.[72]

Conclusion

At the beginning of 1979, David Hamburg appeared before the IOM Council and made a special announcement. He had been reflecting on his three years as head of the IOM. Some Council members had suggested that he serve another five-year term beyond his present one. Hamburg told the Council members that he had decided not to do so. He would complete his term and then leave the IOM. Hamburg announced his intention to establish "valuable, long-term directions for the Institute" during the remainder of his term. In the meantime, the IOM would have plenty of time to search for his successor.[73] Council members greeted Hamburg's decision with genuine disappointment; they wanted him to stay, continue his inspirational leadership, and finish the job of establishing the IOM as an independent organization that was not afraid to engage questions that were close to the surface of the nation's political life.

David Hamburg left office only a month before the presidential election that would chase Jimmy Carter from town. Their simultaneous departures marked the start of a new and very different era in IOM history. At the end of Hamburg's tenure, Philip Handler wrote him a graceful letter that eloquently captured his contributions to the IOM:

> Under your leadership, the Institute of Medicine has been brought to maturity. It has earned a place in the Washington scene and become the instrument to which we aspired when it was created. Our country has yet a long way to go in the development of an accepted philosophy which will enable us to frame a consistent national health policy. Thanks to you, I am confident that the Institute of Medicine will make cardinal contributions to that process. We have enjoyed your boundless good humor, basked in the warmth of your compassion, and been stimulated by the keenness of your intellect. All of us are richer for your stay among us.[74]

Notes

1. Clipping with White House note and Lawrence Meyer, "Panel Urges Curb on Use of Costly X-Ray Device," *Washington Post,* May 3, 1977, in Yordy Files, Accession 91-051, Institute of Medicine (IOM) Records, National Academy of Sciences (NAS) Archives.

2. "Steps Taken in the 1975 Search," Presidential Search Committee Files, 1975–1980, Accession 93-192, IOM Records; Philip Handler to Julius Richmond, June 26, 1975; Richmond to Handler, July 10, 1975; and Handler to David Hamburg, June 26, 1975, all in IOM-1975 File, NAS Records, NAS Archives.

3. David Hamburg to Philip Handler, July 19, 1975, NAS Records.

4. NAS Press Release, October 30, 1975, NAS Records; Curriculum Vitae for David Hamburg, March 1975, IOM Records.

5. "Remarks by David A. Hamburg on the Occasion of His Inauguration as President of the Institute of Medicine, November 6, 1975," Yordy Files, Accession 91-051, IOM Records.

6. IOM Council Meeting, Minutes, November 6 and 7, 1975, IOM Records.

7. Institute of Medicine, *Medicare–Medicaid Reimbursement Policies,* (Washington, D.C.: National Academy of Sciences, 1976), pp. 5, 9–10.

8. IOM Council Meeting, Minutes, March 18, 1976, IOM Records.

9. See, for example, the discussion in IOM Council Meeting, Minutes, May 20, 1976, IOM Records.

10. The discussion of the Lewin report that follows is drawn from "Summary of Lewin Report on Major Issues Confronting the New IOM President," IOM Records, and "Notes on Report to Institute of Medicine Staff by Larry Lewin, March 19, 1976," both in Yordy Files, Accession 91-051, IOM Records.

11. Jim Lewis to Program Director's Group, April 2, 1976, and Lewis to Program Director's Group, April 26, 1976, both in IOM Records.

12. Dorothy Rice to David Hamburg, July 1, 1976, Yordy Files, Accession 91-051, IOM Records.

13. Ernest W. Saward, M.D., to Dr. David Hamburg, June 28, 1976, Yordy Files, Accession 91-051, IOM Records.

14. Robert Petersdorf to David Hamburg, June 18, 1976, Yordy Files, Accession 91-051, IOM Records.

15. IOM Council Meeting, Minutes, July 15, 1976, IOM Records.

16. IOM Program Committee Meeting, Minutes, July 22, 1976, Program Committee Files, Accession 81-006, IOM Records.

17. IOM Council Meeting, Minutes, July 29–30, 1976, IOM Records.

18. David Hamburg, "Institute of Medicine, 1976, Current Status and Future Prospects," IOM Records.

19. "Institute of Medicine Divisional Structure," Yordy Files—Second Series, Accession 94-111, IOM Records.

20. IOM Program Committee Meeting, Minutes, February 1977, April 1977, and July 1977, all in Program Committee Files, Accession 81-006, IOM Records.

21. W.A. Lybrand, IOM Staff, to Division Directors, June 22, 1979, Accession 81-006, IOM Records.

22. David Hamburg to IOM Staff, December 22, 1978, Yordy Files, Accession 91-051, IOM Records.

23. IOM Council Meeting, Minutes, November 1976 and January 1978, IOM Records.

24. Institute of Medicine, Division of Health Manpower and Resources Development, *A Manpower Policy for Primary Health Care: A Report of a Study* (Washington, D.C.: National Academy of Sciences, 1978), pp. 3–9.

25. Richard M. Schleffer, Neil Weisfield, and E. Harvey Estes, "A Manpower Policy for Primary Health Care," *New England Journal of Medicine,* 298 (May 11, 1978), pp. 1058–1062.

26. Summary of "Evaluation of Health Care Quality Assurance Programs," July 5, 1977, Yordy Files—Second Series, Accession 94-111, IOM Records; Institute of Medicine, *Assessing Quality in Health Care: An Evaluation* (Washington, D.C.: National Academy of Sciences, 1976).

27. This account is based on Guido Calabresi to John Hogness, July 29, 1977, and other materials, including summaries of steering committees and grant proposals, in Medical Injury Compensation Study Files, Accession 80-010, IOM Records.

28. Institute of Medicine, *Beyond Malpractice: Compensation for Medical Injuries* (Washington, D.C.: National Academy of Sciences, 1978), p. 65.

29. IOM Council Meeting, Minutes, November 1976 and January 1977, IOM Records.

30. Elena Nightingale to Bernard Greenberg, January 27, 1977, Polio Vaccine Study Files, Accession 91-007, IOM Records.

31. "Evaluation of Poliomyelitis Vaccines: Report of the Committee for the Study of Poliomyelitis Vaccines," Polio Vaccine Study Files, Accession 81-007, IOM Records.

32. Elena Nightingale to the Committee for the Study of Poliomyelitis Vaccines, October 7, 1977, Polio Vaccine Study Files, Accession 84-007, IOM Records.

33. "An Appraisal of Computed Tomography," in "Activities of the Division of Health Care Services," July 1977, Yordy Files—Second Series, Accession 94-111, IOM Records; IOM Council Meeting, Minutes, January 1977, IOM Records.

34. Institute of Medicine, *Controlling the Supply of Hospital Beds* (Washington, D.C.: National Academy of Sciences, 1976).

35. IOM Council Meeting, Minutes, March 23, 1977, IOM Records.

36. IOM Council Meeting, Minutes, March 23, 1977, and May 18, 1977, IOM Records; Edward M. Kennedy, Paul Rogers, Richard Schweiker, Tim Lee Carter, and Jacob Javits to David Hamburg, January 28, 1977, Yordy Files, Accession 91-051, IOM Records.

37. IOM Council Meeting, Minutes, May 17, 1978, IOM Records.

38. IOM Council Meeting, Minutes, July 20, 1977, IOM Records.

39. IOM Council Meeting, Minutes, September 28, 1977, IOM Records.

40. Institute of Medicine, *Strengthening U.S. Programs to Improve Health in Developing Countries* (Washington, D.C.: National Academy of Sciences, 1978), pp. ix, ES7, ES18.

41. Steering Committee Meeting, Minutes, November 29, 1979, Files of Health and Behavior Study, Accession 82-076, IOM Records.

42. Institute of Medicine, *Health and Behavior: A Research Agenda, Interim Report Number 1, Smoking and Behavior* (Washington, D.C.: National Academy of Sciences, 1980), p. 11.

43. Institute of Medicine, *Health and Behavior: A Research Agenda, Interim Report Number 2, Combining Psychosocial and Drug Therapy* (Washington, D.C.: National Academy of Sciences, 1981), p. 8.

44. Institute of Medicine, *Health and Behavior: A Research Agenda, Interim Report Number 6, Behavior, Health Risks, and Social Disadvantage* (Washington, D.C.: National Academy of Sciences, 1982), p. 5.

45. Institute of Medicine, *Health and Behavior: A Research Agenda Interim Report Number 5, Health, Behavior, and Aging* (Washington, D.C.: National Academy of Sciences, 1981), p. 1.

46. Institute of Medicine, *Health and Behavior: A Research Agenda, Interim Report Number 4, Infants at Risk for Developmental Dysfunction* (Washington, D.C.: National Academy of Sciences, 1982), p. 3. Other conferences covered the subjects of bereavement and sudden cardiac death.

47. Institute of Medicine, David A. Hamburg, Glen R. Elliott, and Delores L. Parron, eds., *Health and Behavior* (Washington, D.C.: National Academy of Sciences, 1982), pp. 2, 5, 7, 16, 18.

48. Fred Robbins to Robert Rubin, M.D., Assistant Secretary of Planning and Evaluation, Health, Education, and Welfare, November 5, 1981, Research on Stress and Human Behavior Files, Accession 82-077, IOM Records.

49. Steering Committee Meeting, Minutes, November 26, 1979, "Research on Stress in Health and Disease," Accession 82-077, IOM Records; Institute of Medicine, *Research on Stress and Human Health* (Washington, D.C.: National Academy of Sciences, 1981), p. xi.

50. Glen Elliott, Study Director, and Fred Solomon, Division Director, to Steering Committee Members, August 7, 1981, and Elliott and Solomon to Fred Robbins, August 6, 1981, both in Accession 82-077, IOM Records; *Research on Stress and Human Health*, pp. 6, 20.

51. Institute of Medicine, *Annual Report,* Year Ended June 30, 1979 (Washington, D.C.: National Academy Press, 1980).

52. Chester W. Douglass to David Hamburg, September 7, 1976, and "The Possible Inclusion of Dental Care Under National Health Insurance," December 10, 1976, both in Dental Care Study Files, Accession 81-064, IOM Records.

53. Steering Committee Meeting, Minutes, May 25 and 26, 1978, Study of Dentistry in National Health Insurance, Accession 81-064, IOM Records.

54. Steering Committee Meeting, Minutes, Study of Dentistry in National Health Insurance, September 13, 1978, and Walter McNerney to Lester Breslow, July 25, 1978, both in Accession 81-064, IOM Records.

55. David Hamburg to Ben Barker, September 12, 1979; Karl Yordy to Lou Cranford, December 19, 1978; Linda Demlo to Yordy, March 27, 1979; and "Public Policy Options for Better Dental Health," December 1980, all in Accession 81-064, IOM Records.

56. David Hamburg to Anne R. Somers, October 4, 1978, "US Plans a Major Health Report," *New York Times,* July 13, 1978, and other materials in Surgeon General's Report Files, Accession 86-064-06, IOM Records.

57. David Hamburg to Robert F. Murray, Jr., August 27, 1979, Accession 86-064-06, IOM Records.

58. Victor Cohen, "New Health Report Stresses Prevention Through Diet, Habits," *Washington Post,* July 29, 1979, p. 1; Joseph Califano, *Governing America: An Insider's Report from the White House and the Cabinet* (New York: Simon and Schuster, 1981).

59. Karl Yordy to David Hamburg, April 1, 1976; Yordy to Patrick DeLeon, Senate Commerce Committee, April 29, 1976; and Senator Daniel Inouye to David Hamburg, July 13, 1977, all in Program Committee Files, Accession 94-111, IOM Records.

60. David Hamburg to Senator Daniel Inouye, June 30, 1977, Program Committee Files, Accession 94-111, IOM Records.

61. See materials in the Palliative Care Conference File, Accession 94-111, IOM Records.

62. Karl Yordy to Robert Sparks, Program Director, W. K. Kellogg Foundation, July 30, 1979, Funding Files, Accession 91-045, IOM Records.

63. Karl Yordy to Donald Fredrickson, February 11, 1975; Walter Unger to Roger Bulger and Karl Yordy, May 14, 1975; John Sawyer to Philip Handler, October 5, 1978; Robert Sparks to John Coleman, October 27, 1978; and George Taber, Director, Richard King Mellon Charitable Trusts, to David Hamburg, June 19, 1979, all in Funding Files, Accession 91-045, IOM Records.

64. Eli Evans to "Lee" Schorr, March 24, 1978; David Hamburg to Evans, January 30, 1979; and Evans to Philip Handler, June 12, 1979, all in Funding Files, Accession 91-045, IOM Records.

65. IOM Council Meeting, Minutes, July 1979, IOM Records.

66. IOM Council Meeting, Minutes, May 17, 1978, IOM Records.

67. John Hogness to Wallace Waterfall, June 20, 1990, IOM Records.

68. Philip Handler to Dr. Courtland Perkins and Dr. David Hamburg, March 16, 1978, Yordy Files, Accession 91-051, IOM Records.

69. Courtland D. Perkins and David Hamburg to Philip Handler, July 17, 1978, Accession 91-051, IOM Records.

70. IOM Council Meeting, Minutes, May 29, 1980, IOM Records.

71. Executive Session of the IOM Council, Minutes, July 16, 1980, IOM Records.

72. In David Hamburg's original scheme there had been six divisions, but one had been "deactivated" during fiscal year 1979. IOM Council Meeting, Minutes, November 28, 1979, IOM Records.

73. IOM Council Meeting, Minutes, January 1979, IOM Records.

74. Philip Handler to David Hamburg, September 22, 1980, NAS Records.

4

Fred Robbins and the Sproull Report

On March 19, 1980, David Hamburg informed the Institute of Medicine Council that Frederick C. Robbins, dean of the Case Western Reserve University School of Medicine, had accepted the offer of the National Academy of Sciences (NAS) to become the IOM's next president. The appointment, he said, had been welcomed "enthusiastically" by all and constituted the "most positive response" that he had ever seen to a change in organizational leadership.[1] The leaders of the IOM realized that Robbins would need all the goodwill he could muster. If nothing else, raising funds would be difficult in an economy that was suffering from stagnant employment and sharply rising prices. A week later, Theodore Cooper, head of the IOM Program Committee and former Assistant Secretary for Health in the Ford administration, elaborated on the problems that Robbins would have to solve. Cooper said that the IOM could not let the "somewhat depressing overall economic picture in the country" lower the level of creativity at the IOM, nor could it let "funding sources control its program" during a period of fiscal constraint.[2]

As things turned out, Robbins did face many trials. Some of these challenges came from external conditions, such as the state of the economy and the change in administration from Democratic to Republican. Others stemmed from internal conditions, such as the change of administration at the National Academy of Sciences. The Robbins era culminated at the end of 1984 with a report from an NAS committee that seriously questioned whether the IOM should continue to exist as an entity of working members who engaged questions of social importance. The report implied that the IOM designed by Walsh McDermott should come to an end and a new National Academy of Medicine should take its place. Ultimately, Robbins managed both to expand upon the work of David Hamburg and to reaffirm the concept of the IOM. Child health became an important area of concern for the IOM during the Robbins years; so did studies related to vaccine supply and medical technology. Still, there were always distractions. Just as organizational questions had preoccupied the Board on Medicine, so structural matters consumed the time of Fred Robbins. Although the IOM survived the challenges that arose between 1980 and 1985, the ordeal sapped some of the organization's energy that could otherwise have gone into its expansion.

Searching for Frederick Robbins

In picking David Hamburg's successor, the IOM knew it had a reasonable amount of lead time. Hamburg announced his resignation at the beginning of 1979, and a new IOM president would not have to be in place until the fall of 1980. In March 1979, the IOM created a search committee, composed of members of the Executive Committee, two other members of the IOM Council, and one former Council member. This group selected William Danforth, the chancellor of Washington University in St. Louis, to be its chairman. At its first meeting, the search committee elaborated on the desirable qualities for an IOM president. Some of the characteristics, such as "leadership ability" or a "knack for institution building," were self-evident, but others revealed a great deal about the culture of the IOM. Membership in the National Academy of Sciences was "desirable but not essential"; an M.D. degree was "preferable but not decisive." The committee noted that a "research scientist" might be acceptable if he or she had "good rapport with" and "recognized stature in" the health professional community, implying that it was unlikely the search committee would select, for example, an economist. The IOM was about medicine and would, in all likelihood, be lead by a medical doctor, preferably one who knew something about health policy, who had gained an international perspective on the issues, and who had an "appreciation of the need for constant attention to fund-raising."[3]

The search committee looked for help from two individuals. One was Walsh McDermott, who had been active in all of the previous searches and was the resident expert on the IOM's original mission. The other was Julius Richmond, who had chaired the previous search, which was widely regarded as a great success, and who was in touch with the latest developments in health policy through his government post as Assistant Secretary of Health. The committee's choice of advisers reflected its desire to keep the selection process within the IOM family and not to broaden the search much beyond the IOM and the world of academic medicine. The committee decided, for example, not to advertise the position in a widely circulated publication such as the *New York Times*. Instead of putting an ad in a newspaper or journal, William Danforth solicited nominations from medical school deans, the presidents of 50 colleges, and IOM members.[4]

Despite the closely held nature of the process, the search yielded more than 140 nominations. The search committee interviewed eight candidates and submitted a final list of four to the IOM Council. By the end of 1979, the search committee and the Council had put Frederick Robbins at the top of the list. "Fred Robbins is clearly our first choice," William Danforth told NAS President Philip Handler, "we recommend that all efforts be made to secure his services."[5]

Robbins's career path resembled that of other IOM presidents. After receiving his undergraduate degree at the University of Missouri, he went first to the University of Missouri Medical School and then to Harvard, where he completed his M.D. During the war, Robbins did important laboratory work for the army, identifying the agent that caused a certain form of pneumonia. After the war, Robbins returned to Harvard and continued his work as a pediatrician with a strong research interest in infectious diseases, joining John F. Enders' laboratory and participating in the studies of viruses that caused mumps and polio. The work that Enders, Thomas Weller (who was a charter member of the Institute of Medicine), and Robbins did in cultivating the polio virus earned them the Nobel Prize in medicine in 1954. By this time, Robbins had already left Harvard to become director of pediatrics and contagious diseases at Cleveland's City Hospital and professor of pediatrics at Case Western Reserve University. News of Robbins's Nobel Prize caused quite a stir in Cleveland, although Robbins's wife Alice claimed to be unimpressed, because her father, as it turned out, had also won a Nobel Prize.[6]

Robbins spent the bulk of his career at Case Western. In 1966, he became dean of the Medical School and succeeded in making this school one of the most exciting places to study medicine in the nation. As dean, Robbins fostered the creation of new departments of Community Health and Family Medicine, and he made sure that students received a solid grounding in primary care. Presiding over turbulent times in a calm and unflappable manner, he was, according to his Case Western colleagues, a capable leader, willing to compromise, rather than a firebrand or an innovator. "I'm not going to revolutionize things," he told a reporter at the time of his appointment as IOM president. He sought at first simply to continue the program that David Hamburg had begun.[7]

Robbins was already an IOM insider. Like David Hamburg and Donald Fredrickson at the time of their selection as IOM president, he had served on the IOM Council. Unlike these two, Robbins had also been in residence at the IOM as a senior scholar. Taking a sabbatical during 1977–1978, he came to the IOM and chaired the steering committee for the studies of saccharin and food safety policy. As a member of the IOM's Executive Committee, Robbins might be thought of as part of the Hamburg administration, in a manner analogous to a federal cabinet officer. With perfect justification, Hamburg could tell the members of the IOM that "Fred has been deeply committed to the tasks of the Institute." His accomplishments as physician, scientist, educator, administrator, and government adviser matched "the breadth of the Institute's mission."[8]

Robbins was accomplished, in a sense that set him apart from those who had already served as IOM presidents. At the time of his appointment, Robbins was 63, only three years away from being designated a senior member. He had already been a dean, already become a member of the National Academy of Sciences, and already won the Nobel Prize. Walsh McDermott hoped that the IOM would be led by young and vigorous men and that its ranks would constantly be replenished by younger men. Hogness, Fredrickson, and Hamburg were all young men on the rise, and each went on to greater prominence after having served as IOM president. Hogness became a university president; Fredrickson served as head of the National Institutes of Health (NIH) and head of a foundation; Hamburg took a job at Harvard and then became the head of the Carnegie Corporation. It was unlikely that Robbins, who would be 68 at the end of his term, would follow a similar path. Unlike the personable and energetic Hogness, the intensely intellectual Fredrickson, or the charismatic Hamburg, Fred Robbins, an affable and highly competent man, would maintain what others had created. From the very beginning, he faced an uphill fight. The first thing that NAS President Philip Handler told him was that Handler should never have allowed the IOM to get started.

The 1980 annual meeting, held in the middle of October, marked the formal beginning of the Robbins era. The program that year featured a look at health in the new decade and included speeches by Robert Wood Johnson Foundation head David Rogers, Nobel laureate geneticist Joshua Lederberg, future Nobel laureate economist Kenneth Arrow, and distinguished jurist David Bazelon. Much of the program was celebratory in nature. On the first day, the Institute held a special symposium in honor of David Hamburg in which five distinguished scientists gave papers on "Adaption, Stress, and Coping." The inauguration of Frederick Robbins came on the second day. The audience heard a message from John Enders who reminisced about Robbins's days at Harvard and a speech by Philip Handler on Robbins's scientific and medical contributions.[9]

Backlist of Projects

As Robbins settled in as head of the IOM, he administered a series of projects that began in the Hamburg era and came to a conclusion during his presidency. The situation resembled that of the new head of a Hollywood studio, who had first to market his predecessor's films before he could promote his own. Robbins found a backlist of at least five projects that, in a manner typical of the Institute of Medicine, covered a wide range of subjects.

Some, such as the airline pilot study, were narrow in focus yet of vital interest to the groups involved. The study stemmed from a controversy that had arisen over a Federal Aviation Administration (FAA) rule prohibiting commercial aviation carriers from allowing anyone over 60 to pilot or copilot a plane. As the generation of pilots trained in World War II and Korea aged, they began to feel that the FAA rule was discriminatory and unnecessary. They formed groups such as the Pilots' Rights Association and petitioned Congress to pressure the FAA to change the rule. Congress responded, as it did in many controversial situations, by calling for a study of the matter.[10] At the beginning of 1980, Donald Fredrickson, the head of NIH, contacted David Hamburg about having the IOM do the congressionally mandated study. Although the IOM Council reacted with enthusiasm, it took many months for federal officials to decide if they wanted to enter into a contract with the IOM or whether they preferred to do the study in-house. Not until June was a contract signed and a study group appointed. The IOM panel would confine itself to objective medical findings on the subject of pilot performance and age, and the National Institute of Aging would use the data as part of its formal response to Congress.[11]

The IOM decided to form a series of task forces to examine the effects of aging on various bodily systems, for example, one that focused on the cardiovascular system. The group made site visits to places such as the American Airlines Flight Academy in Dallas, Texas, where it hoped to acquire a sense of the physical skills involved in flying a plane. The final report, a product of the IOM's Division of Health Sciences Policy and a steering committee headed by Robert F. Murray, chief of the Division of Medical Genetics at Howard, appeared in 1981, about five months after Fred Robbins's inauguration.

The report was decidedly tentative in tone. On the one hand, the group noted that for significant acute events, such as a heart attack or stroke, age 60 did not mark the beginning of a period of special risk. On the other hand, the group concluded that "subtle changes that may adversely affect pilot performance" increased with age. A pilot's skills deteriorated with age, yet there was great variation among individuals in any particular age group. The committee implied that tests of individual acuity should be developed that could take the place of a blanket exclusionary rule. In the end then, the report called, as did nearly every IOM report, for further research that would make such tests possible.[12]

Another study on the backlist concerned nursing education and marked a throwback to the large data-gathering studies that had been done by Ruth Hanft. Just as the IOM was asked to investigate how the federal government should subsidize medical education at the beginning of the 1970s, so it received a similar request to investigate nursing education at the end of the decade. In the Nurse Training Amendments, passed in 1979, Congress sought the IOM's help in resolving a controversy over whether there should be continued federal support of nursing education. Although the study would not involve the data-gathering efforts that had marked the earlier study supervised by Hanft, it was nonetheless a large and ambitious undertaking that was scheduled to take two years to complete.[13]

The Institute of Medicine asked Arthur Hess to head the steering committee that included some truly distinguished practitioners in the field of health services research. Hess was a veteran bureaucrat from the Social Security Administration who had helped launch the Medicare program. Other members included Otis Bowen, a medical doctor from Indiana who later became governor of Indiana and Secretary of Health and Human Services; Stuart Altman, dean of the Heller School of Social Welfare at Brandeis; Saul Farber, head of the Department of Medicine at New York University; John Thompson, a professor of nursing education at Yale; Isabel Sawhill, a specialist in manpower policy at the Urban Institute; and Linda Aiken, a nurse with a Ph.D. in sociology who worked for the Robert Wood Johnson Foundation. This panel plunged into the complexities of federal subsidies for nursing education.

The tone of the 1983 report, a product of the IOM's Division of Health Care Services, revealed some of the differences that had occurred over the course of a decade in the field of social policy. The nursing report, unlike the IOM's earlier report on federal support for medical education, did not reflexively call for more federal spending or regulation. On the contrary, the group concluded that "no specific federal support is needed to increase the overall supply of registered nurses, because estimates indicate that the aggregate supply and demand for generalist nurses will be in reasonable balance during this decade." Despite this finding, the study group had no intention of abandoning the IOM's earlier efforts at social reform, arguing, for example, that states and private employers should facilitate the actions of nurses who wished to upgrade their skills, particularly those nurses who wished to pursue clinical, rather than administrative, careers. In a similar vein, the group believed that the federal government should cosponsor, with states and private foundations, demonstration projects designed to alleviate nursing shortages in medically underserved areas and that the federal government should institute a "competitive" program to provide scholarships for members of minority and ethnic groups who wanted to be educated as nurses. Nurse practitioners also received the study group's endorsement, as in the recommendation that the federal government should continue to support the training of nurse practitioners who were needed in medically underserved areas and in programs caring for the elderly. Words such as "competitive" and public–private or federal–state "partnerships" reflected the new conservative tone of the era. One could no longer simply assume that health professions education was a public good deserving of federal support or that an increase in the number of health care practitioners was socially desirable. At the same time, the report made it clear that the study group and, one might infer, the Institute of Medicine believed in a strong federal presence in the field of health manpower.[14]

The government's proper role in maintaining an adequate supply of doctors also figured in a study the IOM carried out for the Department of Defense on graduate medical education in the military. A doctors' draft, meant to handle the contingencies of the Vietnam War, ended in June 1973. Without a ready supply of doctors, who were needed to provide care to servicemen stationed around the world and to be prepared for the onset of another war, all three branches of the military faced a crisis. One way around this problem was to pay for a person's medical education in return for a period of military service. Another way was to allow a medical doctor to take his or her internship and residency while on active duty. By the late 1970s, a substantial number of active-duty military physicians were interns, residents, or fellows, and the Department of Defense ordered the services to limit the fraction of military physicians in graduate medical education to not more than 20 percent of their authorized physician strength. All three of the uniformed services protested this ruling. So the Assistant Secretary of Defense for Medical Affairs asked the IOM to undertake a study of graduate education in the military services.[15]

The IOM complied. Using its remarkable ability to assemble experts in a given area of medical policy, the IOM put together a strong steering committee. Leonard Cronkhite, who was both president of the Medical College of Wisconsin and a major general in the U.S. Army Reserves, agreed to chair. The group decided that the policy question of how large the Uniformed Services University of the Health Sciences, which was essentially a military medical school, should be could not be "answered objectively and unambiguously." Nonetheless, the group recommended that the Department of Defense withdraw its directive to limit the percentage of military physicians in graduate medical education assignments to 20 percent. Instead, each military medical department should be permitted to adjust its graduate medical education programs "to meet the changing manpower circumstances and requirements that it faces." In order to do this, the Department of Defense should produce a set of "planning guidelines, programs, and personnel policies" for each of the services.[16] In this manner, the report, done quickly and with rather cursory analysis, contained a variant on the usual "more research is needed" recommendation. It also showed just how reluctant the members of the IOM steering committee were to part with an important federal subsidy to medical education. For many influential IOM members, including every one of its presidents, military service had been an important part of their medical and research training. Of course, they had had little choice about the matter, and their service sometimes stretched into a long period that interfered with their careers. These doctors wanted future generations of physicians to enjoy the opportunities that military service provided, without being disadvantaged in comparison to their civilian peers.

Differences in generational outlook also figured in another of the IOM studies that Fred Robbins inherited from David Hamburg. This study, on the health-related effects on marijuana use, reflected the IOM's role as an arbiter of public health. The National Institutes of Health initiated the request for the study, perhaps because federal officials understood just how controversial a topic marijuana use was and wished to have it investigated by an external authority.[17] For the IOM, the study resembled the one it had done on the health effects of legalizing abortion. Although in both cases the subject was controversial, the IOM restricted its investigations to the health aspects, not the morality, of the subject under investigation. In both cases, the IOM discovered that the subject was so emotional that it was difficult to conduct a dispassionate inquiry. In both cases, the IOM faced a dissonance between the ambiguity of the data and the almost religious certainty of people on opposing sides of the issue.

The study fell into the domain of the Health Sciences Division at the IOM. Enriqueta Bond, a talented scientist who would become an increasingly important part of the IOM's history, did much of the staff work for the project. She had arrived at the Academy complex as a study director in the Division of Medical Sciences of the National Research Council (NRC). IOM staff member Elena Nightingale recruited her to come over to the IOM and work on studies related to genetic screening in the workplace. According to Bond's later recollection, "the IOM was the new kid on the block doing innovative things and having substantial impact in the policy world."

On the marijuana study, Arnold Relman, editor of the *New England Journal of Medicine*, played a prominent role as chairman of a large steering committee that was divided into many subpanels that examined such aspects of the problem as reproductive and fetal issues and behavioral and psychosocial issues. When Relman met with the group for the first time on April 15, 1981, he told members that their task was to write a report that could "be used a guide to policy but which itself is not required to suggest policy—thank God." He noted that the group operated in "the context of a very polarized social and political scene" and that its work would be "watched and prodded and . . . criticized by many contending forces." He suggested that the members be "totally indifferent to these social and political forces and just do our jobs as scientists and physicians and evaluators, dispassionate evaluators of the evidence."[18]

Even as Relman spoke, the IOM began to receive scores of letters from grassroots proponents and opponents of marijuana use. For example, Robert L. Mitzenheim urged that marijuana be used by people with glaucoma or undergoing chemotherapy. "After seeing people go through these things without even the chance to try the THC [an abbreviation of the chemical name for the active ingredient in marijuana], my feelings have become stronger in opening new doors for them; I personally would use it, law or no law," he wrote. Thomas E. Campbell reported that he smoked marijuana three or four evenings a week and found that it relaxed him and enabled him to sleep. With regular marijuana use, Campbell said, he could reduce his alcohol consumption below his previous level of six drinks per day and his use of tobacco from his previous pack-a-day habit. The steering committee also received mail from a Martha Stone, who offered testimony of how her child's marijuana use produced four years of "absolute chaos in our home." A writer from Celeste, Texas, told the lamentable tale of her nephew Rick who was sitting in jail awaiting trial. When he was in elementary school, he was the ideal child, "smart, handsome, active, enjoyable." Then he discovered girls and marijuana, and paradise was lost. After that, "instead of bringing home those straight A report cards, he brought home misery."[19]

The IOM ignored these letters and probed the facts. Relman pressed steering committee members who brought him anecdotes on the detrimental effects of marijuana to bring him the facts. When one member told him that marijuana could produce an "acute brain syndrome," Relman asked if the phenomenon was well documented in the literature. Told that it was not, Relman replied, "We have to be very careful. We cannot cite our own clinical experience in this."[20] The reality was, however, that facts could take the group only so far. As a steering committee member pointed out, "the ultimate decision about the question may be nonscientific, maybe more a political, social, cultural, moral kind of thing." One member went so far as to suggest that the IOM get some "young people to read some of the report and see if it has verisimilitude."[21]

The final report, which came out at the end of 1981, took a very cautious approach to the subject. It pointed out, for example, that smoking marijuana produced "acute changes in the heart and circulation that are characteristic of stress" but concluded there was no evidence "to indicate a permanently deleterious effect on the normal cardiovascular system occurs." Marijuana produced acute effects on the brain, but there was simply not enough evidence to determine "whether prolonged use of marijuana causes permanent changes in the nervous system or sustained impairment of brain function and behavior in human beings." Although marijuana might possibly be useful in the treatment of glaucoma, the control of nausea and vomiting, and the treatment of asthma, the group could not say this for sure and thought that other, already approved therapeutic agents made more biological sense than THC. The antidote to all of this ambiguity was more research—the IOM's universal antidote. "The explanation for all these unanswered questions is insufficient research," the report noted.[22]

For all of its caution, the report could be read as an indictment of marijuana use. "Our major conclusion," the steering committee wrote, "is that what little we know for certain about the effects of marijuana on human health—and all that we have reason to suspect—justify serious national concern."[23] The committee worried particularly about the effects of marijuana on child development. In these ways, the marijuana study resembled the abortion study. In both cases, the fundamental beliefs of the steering committee were clear: in the one instance, prochoice, and in the other, antimarijuana. Both reports left the impression that further research would buttress these basic beliefs.

Not everything that the IOM studied in the areas of public health and health sciences concerned subjects of long-standing duration. Just as Fred Robbins took over from David Hamburg, the IOM became involved in the emerging concern over toxic shock syndrome. This term first came into currency in 1978 and drew a large amount of attention in the spring and summer of 1980. At that time, epidemiologists discovered a link between cases of toxic shock, which could produce dangerous fevers and even death, and young women's use of tampons. A particular brand of tampon, known as Rely and manufactured by Procter and Gamble, appeared to produce a disproportionate number of cases. In response, Procter and Gamble withdrew Rely from the market in September 1980. A few weeks later, Johnson and Johnson, also a manufacturer of tampons, approached the IOM about doing a study of toxic shock syndrome. Very aware of the need to protect the IOM's reputation for objectivity, Fred Robbins told Johnson and Johnson that the IOM would like to do the study but would require a broader base of funding. In the end, four government agencies, Johnson and Johnson, and Procter and Gamble all helped to finance the project. It proved to be an interesting collaboration between the IOM and private industry. When the steering committee met for a second time on September 21, 1981, for example, it heard from both of the manufacturers under an elaborate set of guidelines. The representatives of one company agreed to leave the room when representatives of the other spoke, in an effort to protect proprietary information.[24]

The report on toxic shock syndrome was highly technical in tone. The steering committee, chaired by Sheldon Wolff, head of the Department of Medicine at Tufts and physician-in-chief of the New England Medical Center, noted that cases of toxic shock were rare but severe. It therefore seemed prudent to give women enough information to allow them to make informed decisions about tampon use. At the same time, the committee felt that the available information indicated that women who had had toxic shock syndrome, who were postpartum, or both should not use tampons. Women between 15 and 24 years of age who used tampons needed to understand that they were at higher risk than older women, and all women should minimize their use of high-absorbency tampons of the Rely type.[25]

The toxic shock syndrome project showed that the disease environment changed with time. The very conditions of modern life produced new health complications. In this particular case, the public needed an organization it could trust to judge the safety of a product used by a high percentage of women. The IOM, with its ability to assemble experts relatively quickly and its complete independence from large companies such as Johnson and Johnson, met this need.

The Ebert Report

Circumstances forced Fred Robbins to spend a great deal of time removed from the sorts of scientific considerations that informed the marijuana and toxic shock studies and to be engaged instead in practical matters of governance. A case in point concerns the rules for membership in the IOM. When Robbins took over at the end of 1980, the organization faced a major decision. The initial group of members, selected by Robert J. Glaser and members of the Board on Medicine, joined the organization in 1971. The rules allowed them to serve two five-year terms before they became senior members. Senior members could take full part in the IOM's activities with two significant exceptions. They could neither vote in the final membership elections nor serve on the IOM Council. As many as 76 members would complete their 10 years of active service at the end of 1981. As the date approached, the IOM Council began to reconsider whether it wanted to risk losing the services of so many of its experienced members.[26]

The answer was no. Instead of following the Walsh McDermott dictum that the IOM always be in the hands of young and energetic members, the IOM chose to make membership almost a lifetime condition. On September 16, 1981, the IOM Council voted to abolish the two-term limit. A man who was elected at age 45, for example, could continue as an active member until he reached age 66, at which time he became a senior member. Before this age, he would receive a letter from the IOM president every five years, asking if he wished to continue in active membership or become a senior member. If he wished to continue, all he had to do was say so. At the same time the IOM Council made this decision, it also decided to raise the ceiling on total membership in an effort to leave room for new members. Before 1981, the number of active members was capped at 400. The Council wanted to abandon the ceiling altogether; IOM members preferred some sort of limit. The compromise was to raise the maximum number of active IOM members to 500 in 1981. In agreeing to these changes, Robbins recognized the risk that the IOM could be dominated by "relatively older members"—in other words, people like himself. He pointed out, however, that because the average age of election to the IOM was 53, the abolition of terms added only three years to an average active member's term.[27]

At the same time the IOM decided, in effect, to conserve its membership, it also took steps to protect its basic structure. At the end of David Hamburg's term, he appointed a task force to examine the IOM's structure and its relationship to the rest of the National Academy of Sciences. William Danforth agreed to chair the group. Even though the task force originated with Hamburg, Fred Robbins essentially took it over, and it became his vehicle for shaping the organization to his liking. When the task force met in the spring of 1981, it examined such questions as the distinctions between active and senior members and the relationship between the IOM and the Assembly of Life Sciences (ALS) in the National Research Council. Robbins decided that the effort should culminate in a special retreat that summer at which the IOM Council could take stock of the organization's progress.[28]

The need for such a review was made more urgent by the fact that the National Academy of Sciences also had a new president. Frank Press, a distinguished physical scientist from the Massachusetts Institute of Technology (MIT), who had served as President Carter's science adviser, took command of the NAS in 1981. He put his vice president, James Ebert, an embryologist from the Carnegie Institution, in charge of reviewing the relationships among the various NAS units. Both Press and Ebert agreed to meet with the IOM Council at the July retreat.[29]

The interim report of the IOM task force on structure formed the basic text for the July retreat. It could be thought of as an affirming document because it validated the IOM's philosophical underpinnings. The report stated that there was no need for a fundamental change in the IOM's organizational structure and that the IOM should continue to examine issues "with intertwined policy and scientific components." Indeed, "a central element in the whole concept of the IOM was the blending of these two domains of study." In order to accomplish this blend, the IOM should seek flexible funds that would give it more freedom to undertake studies at its own discretion. It should also work with the National Research Council to streamline the cumbersome report review process and to eliminate the overlap between the IOM's activities and those of the Assembly of Life Sciences in the NRC. In particular, the report recommended that the NAS consider removing the Division of Medical Sciences (which was part of the ALS) from the NRC and putting it in the IOM. Short of this, the IOM president might serve as the chairman of the Division of

Medical Sciences. The IOM president, burdened with so many responsibilities, should also be allowed to name a vice president to lighten his load.[30]

The IOM Council members agreed with the task force that the IOM should remain true to its original design and mission. As a corollary, the Council advised the IOM to create a resources development committee to develop a plan for raising "an adequate pool of independent, flexible funds" that would meet the Institute's long-range financial needs. The option of appointing a vice president "should remain open." More urgently, the IOM should persuade Frank Press to revise the NAS report review system and should point out to him "as forcefully as possible" the detrimental effects of maintaining so many redundant units within the Academy.[31]

In the fall of 1981, Fred Robbins wrote Frank Press that there was a "significant and disturbing problem of overlap in the program interests of the IOM and the ALS." The program plan, style of operation, and "general scope" of the Division of Medical Sciences "increasingly converge" with the IOM, leading to "friction, 'turf' sensitivities, and confusion" in the minds of funders. After 10 years of existence, the IOM had proved its worth to the NAS, and it "would be unfortunate to compromise the IOM's value by allowing . . . a diminution of its vitality through organizational overlap." The solution was to merge the Division of Medical Sciences into the IOM.[32]

The Ebert report on the organization of the National Research Council appeared in February 1982. It proposed that the NRC be divided into six major units, including one devoted to human health and medicine that would encompass some of the functions of the Assembly of Life Sciences and of the IOM. The unit would be "housed within the Institute of Medicine and overseen by the IOM Council" and would end the overlap between the ALS and the IOM. In other words, the Ebert report suggested that the IOM be given much of what it wanted.[33]

The IOM greeted the Ebert report with enthusiasm. In March, the Council approved the concept of merging IOM and ALS activities and agreed that it should become the governing body of the new commission. As part of its new responsibilities, the Council decided that it should contain more members from the biomedical sciences.[34]

In the end, little substantive change occurred. The National Research Council did get a substantial facelift in the spring of 1982 with the creation of new commissions and boards. What were assemblies in the old organizational scheme became commissions in the new one. Hence, the old Assembly of Life Sciences became the new Commission on Life Sciences. Still, the NAS Council hesitated to confer greater power on the Institute of Medicine, and as a result, the merger of the Division of Medical Sciences and the IOM never took place. Although the Division of Medical Sciences was abolished, the NAS failed to authorize a Commission on Human Health and Medicine. Instead, Fred Robbins, the president of the IOM, became chairman of the Commission of Life Sciences in the summer of 1982.[35]

The resolution of organizational issues, however unsatisfactory, freed the IOM to concentrate on other pressing concerns, such as the criteria for membership and the creation of a realistic program plan. In July 1983, Robbins appointed a special task force, headed by John Hogness, to investigate the membership selection process. Hogness's task force sought to counter the "general perception that although the general level of quality remains high, the quality of some candidates nominated for and elected to membership has been marginal." The solution, the group decided, was to incorporate "a minimum threshold of accomplishment" into the process. Robbins also tried to tighten the process by which the organization planned its activities, although this proved to be an arduous task. In 1981, for example, Irving London, chairman of the advisory committee to the IOM's Division of Health Sciences Policy, reported with some exasperation that attendance at meetings was spotty, enabling a few people to dominate the proceedings. The division did not set its own priorities as much as respond to often highly specific requests from government agencies, with the result that broader studies tended to be neglected. London concluded that the division needed independent funding so as not to be dependent on federal agencies.[36]

Advent of the Reagan Administration

The arrival of the Reagan administration made the need for financial and intellectual independence seem all the more urgent. Even at the end of the Carter administration, the IOM experienced considerable difficulties securing contracts from the federal government. The change in administrations exacerbated the problems, as efforts to work with such agencies as the Health Care Financing Administration indicated.

The Health Care Financing Administration (HCFA), whose purpose was to unite the administration of the Medicare and Medicaid programs, was a creation of Joseph Califano and the Carter administration. It appeared logical that the IOM would work with HCFA as it pursued questions related to the reimbursement of hospitals and physicians and as it handled issues of quality of care.

The IOM sought a cooperative agreement—a mechanism somewhere between a grant and a contract—with HCFA that would assure the agency of ready IOM advice on these matters. It helped that Robert Derzon, the first head of HCFA, had a close relationship with the IOM and later became an IOM visiting fellow. When Leonard Schaeffer, who was Derzon's successor, met with the IOM Council on May 9, 1979, he expressed a strong interest in working with the IOM on physician reimbursement issues. On Schaeffer's last day in office, he signed a cooperative agreement with the IOM, yet nothing came of it. At the end of December 1980, for example, the IOM sent HCFA a proposal for a workshop on physician reimbursement policies under Medicare and Medicaid, but it was clear that HCFA was wary of the proposal. At most, the agency wanted only a forum for physicians "to ventilate their concerns," and it did not wish the IOM to make specific recommendations on changes in physician reimbursement. HCFA officials, it became evident, had no desire to tie the hands of the Reagan administration before it had even begun.[37]

The IOM never did find its niche in the area of hospital and physician reimbursement under Medicare. In 1983, Congress approved a prospective payment system for hospitals that depended on a statistical mechanism known as a "diagnosis-related group." Although the IOM had once made a similar proposal, it played no role in the development of the one that was adopted. After the 1983 legislation, IOM members continued to cite physician reimbursement as a priority and developed an elaborate proposal to evaluate alternative methods of paying physicians. Despite the self-described "intense interest" of IOM members in this subject, HCFA rejected the IOM proposal and awarded instead a contract to a group of Harvard researchers to study what were known as "resource-based relative value scales." These scales provided a means of comparing the services of doctors in different medical specialties—for example, a surgeon and a general practitioner—and served as a basis for changes in physician reimbursement under the Medicare program that Congress made in 1989. Once again, the IOM played no role in the research that led to the change.[38]

Not all of the IOM's overtures to the government were rejected in the Reagan era. The organization achieved notable success in forming cooperative agreements with the Public Health Service (PHS). The parties signed the first such agreement in February 1981. With the funds received, the IOM transformed three of its divisional advisory committees into boards that could conduct independent investigations. The new Board on Health Promotion and Disease Prevention, for example, took the place of the old advisory committee for this IOM division. Support from the PHS was neither certain nor steady, yet it persisted throughout the Reagan administration. In 1982, the Public Health Service could provide only 60 percent of the amount it had given the year before. Federal officials, such as Edward Brandt, the Assistant Secretary for Health, indicated that they would try to increase the amount in the following year. In 1983, the Public Health Service came through with more money, extended the agreement for another year, and made plans to support a fourth division board.[39]

Despite this cordial gesture from the Public Health Service, the fact was that the Reagan administration had a chilling effect on the IOM's development and on the entire National Academy of Sciences complex. In January 1981, for example, the IOM learned that a study of federal research policies that was considered promising would not be funded by the Department of Health and Human Services (HHS) because Richard Schweicker, the new HHS Secretary, had other priorities. The Democratic Congress that left town in 1980 had wanted the IOM to review the state of the art in neurological research. The mixed Congress—Democratic House and Republican Senate—that came to town in 1981 refused to appropriate money for the study. Senator Edward Kennedy (D-Mass.) wanted the IOM to form a Council on Health Services Research, but IOM leaders knew that legislation mandating such a council would never be passed. Robbins believed that without such a mandate, further discussion would be "fruitless." Even with a legislative mandate, it was often difficult to undertake a study. The Omnibus Budget Reconciliation Act of 1981 contained a provision for the IOM to study the implications of the increasing number of physicians. With the administration less than enthusiastic, the study was never funded.[40]

The hostility of the Reagan administration hit the entire NAS hard, particularly after President Carter's science adviser made the transition to the Academy. Many parts of the NRC received fewer contracts to undertake studies, just when the NAS had leased a new building, known as the Joseph Henry Building, from George Washington University. The administration also attempted to make deep cuts in the budgets for the National Institutes of Health, the National Science Foundation, and the other parts of the government that supported scientific and statistical research. This prompted serious concern at the Academy, which sponsored a conference in October 1981 to consider the impact of the administration's prospective budget cuts, and at the IOM, where Council members noted that budget cuts would affect "the entire NIH enterprise" and academic health centers as well. As for support for health services research, it was practically "defunct," according to Johns Hopkins professor of pediatrics and health services research and IOM Council member Barbara Starfield.[41]

This sort of discussion underscored the fact that the IOM, despite its record of serving both Democratic and Republican administrations, often found itself in opposition to the Reagan administration. Officials in the Reagan administration, for their part, often interpreted the IOM's actions as hostile. For example, the IOM offered a number of Democratic officials a setting in which to make the transition from government to other types of work. In this manner, Howard Newman, a former director of HCFA, became an IOM member in residence during 1981. For the IOM, the offer to Newman was similar to the one it had made to Theodore Cooper at the end of the Ford administration and indicated no disapproval of the Reagan administration. To the administration, however, which was intent on making revolutionary changes in the relationship between the government and its citizens, harboring former Democratic officeholders was a suspicious act. At times too, the IOM's advocacy of an agenda that it had built up over the course of a decade created the possibility of conflict with the Reagan administration, as in the desire of some IOM members to issue a statement decrying the lack of access to health care for those unable to pay. The IOM Council squelched this initiative, saying it did not wish to issue a "moral statement" that was not supported by data. The Council did encourage the IOM to do a formal study of ways to facilitate access to health services for those unable to pay for them. Although such a study represented an important priority of the IOM's Program Committee, the NRC governing board rejected it for fear that it "implied subtle criticism of the current administration." Robbins replied that the "Institute strives to be unbiased about political issues." Still, "occasions may arise when it must make recommendations that are contrary to current administration policy." Whatever the imperatives of the moment, the study was never done. A study of the medical consequences of nuclear attack on the United States languished for similar reasons.[42]

Studies During the Reagan Era

Many of the studies that the IOM launched during the Robbins's presidency reflected the influence of the changed ideological climate induced by the Reagan administration. Where before the Institute had studied "health care in a context of civil rights," it now turned to the growth of "for-profit investment in health care." First suggested in 1981, the study was not begun until 1983, when the IOM secured partial funding through the Andrew Mellon and John A. Hartford foundations and through a number of private companies in the hospital management business. As always, however, the IOM took pains to ensure that there was no impropriety in the funding arrangements, working hard to guarantee that the private companies had nothing to do with picking the members of the steering committee or conducting the study. To set the tone, the IOM hoped to interest a "distinguished citizen with no vested interest" in chairing the study committee. After John Dunlop, the Harvard professor and former Secretary of Labor who had briefly served on the Board on Medicine, and Kingman Brewster, the former president of Yale, declined, the IOM settled on Walter McNerney of the Blue Cross Association.[43]

Despite the many financial contributors, the project was run on a shoestring, with the constant need to seek additional funding. At one point, for example, Brad Gray, the project staff director, decided to abandon commissioning two papers on ethical issues for want of money. Roger Bulger, president of the University of Texas Health Science Center at Houston and former IOM staffer, came to Gray's rescue by agreeing to have his institution fund the papers.[44]

The bulk of the committee meetings took place in 1984, during the period in which Ronald Reagan thrashed Walter Mondale in the general election. Discussions revealed the polarized opinions that people brought to the table. Kenneth Platt wondered if profits in health care were "obscene," because they led to discrimination against the poor and produced negative influences on physician behavior. John Bedrosian pointed to the many positive features of the investor-owned segment of the health care industry, such as the creation of new facilities, the development of services, and the investment of millions in underserved communities. Paul Ellwood, the noted authority on health maintenance organizations, predicted that in 10 years there would be only 10 large suppliers of health care, because of the large advantages for-profits enjoyed in access to capital. Clark Havighurst, an expert on health policy and the law, reminded the committee of all the changes that had taken place in the field: competition had become mandatory; advertising was permitted; boycotts were prohibited; and restrictions had been relaxed against the corporate practice of medicine. Above all else, decisionmaking power had shifted from physicians to the purchasers of health care, who wanted to reduce costs through such devices as prospective payment. Such changes were bound to have an effect on the quality of care. Havighurst pointed, in particular, to the problem of access for the "near-poor." He argued that the problems should not be corrected by "reregulating the industry." It was better, Havighurst argued, to give the near-poor increased health purchasing power, perhaps in the form of vouchers.[45]

The discussions held by the steering committee on nonprofit enterprise in health care went to the very heart of the changes taking place in the Reagan era. In the end the group debated whether to focus the report on these changes as, in Walter McNerney's words, a "natural outgrowth of public disillusionment with a regulatory perspective and the inefficiency of nonprofits." In this view, the committee should publicize the way in which investor-owned enterprises "seized the opportunity created by cost-based reimbursement, deregulation, and a maturing market." Others argued that merely to describe the historical process missed its meaning. In this view, the report should emphasize how the growth of for-profit health care undermined the role of the physician as an agent for the patient and how the growth of entrepreneurial activities had a deleterious effect on the "fiduciary" relationship between doctor and patient.[46]

The final report reflected a compromise between the two views. It both described the changes that were taking place in the health care industry and tried to assess some of their consequences. The committee concluded that there was no evidence to indicate that investor-owned hospitals were less costly or more efficient than not-for-profit organizations. At the same time, these types of companies did make "services more readily available" to the people they served. Still, there was enough evidence to suggest that the for-profits provided very limited amounts of uncompensated care, which only served to increase the burdens on the nonprofits. Therefore, although the for-profits eased some problems of access, they worsened others. This was a major motif of the report: deregulation in the health care industry solved some problems but created others. A summary of the report prepared by Brad Gray and Walter McNerney ended on a cautionary note. The nation should feel "uneasy" with the recent changes, not because of the "venality" of potential investors but because "markets need legislative and regulatory support to work, especially for essential human services, in which the needs of those without resources cannot otherwise be met."[47] The bottom line, then, was that the poor needed protections that the for-profits could not supply.

When the IOM report on for-profit enterprise appeared in 1986, it attracted a great deal of attention in the popular press through a common media loop. The IOM used the good offices of *New England Journal of Medicine* editor Arnold Relman to publish a summary of the report in that journal. The summary piece, in turn, attracted the notice of the *Wall Street Journal*, the *New York Times*, and the *Washington Post*. Because the editors of these newspapers saw advance copies of the *New England Journal* article, they published pieces on the report on the date it was formally issued. Once the report had received such play in the print media, it became a legitimate subject for a lengthy segment of the "MacNeil–Lehrer Report" on public television that featured a debate between two members of the study committee. Once members of Congress read about the report and saw it mentioned on television, they asked their staffs to request copies. In these ways, the report reached the public and heightened the visibility of the IOM.[48]

Not everything that the IOM did in the Reagan era involved the sort of implicit criticism of the nation's direction that was contained in the for-profit report. A good example of an effective, if somewhat troubled, collaboration came in a study of the organizational structure of the National Institutes of Health that was released in 1984. In 1982, Congress, faced with demands that additional institutes be created, included a request for IOM to study the matter in the Health Research Act. Although the bill never became law, Secretary of Health and Human Services Margaret Heckler requested that the IOM conduct the study anyway. She wanted to understand how the present NIH structure had come to be and how it might be changed. The IOM put James Ebert, the biologist who had recently investigated the structure of the National Research Council, in charge of a steering committee that included, among others, the president of Purdue University, a Nobel Prize-winning biochemist from Vanderbilt, and a leading student of public administration from Harvard's Kennedy School. In addition to hiring a six-person staff for the project, the IOM also commissioned historical case studies of the development of each NIH institute.[49]

The final report was a model of brevity. The IOM published much of the data that had been collected on the National Institutes of Health as appendixes. Recommendations were contained in a 41-page report that also detailed the phenomenal growth of the National Institutes of Health. Between 1943 and 1984, annual NIH expenditures had increased from $1.3 million to $4.5 billion, and staff from 1,000 to 16,000 employees. Even when the rate of inflation was taken into account, the NIH still managed to increase its expenditures at an annual rate of 24 percent between 1943 and 1968. Although the rate had slowed somewhat after that time, the NIH, as the report noted, continued to receive close attention and strong financial support from Congress. Because of its success, the NIH was constantly subject to demands to create new institutes. At the time of the report, advocates were pressing Congress for new institutes in the fields of nursing and arthritis. Ebert's committee recommended that the pressure to create new institutes be resisted. NIH had reached a point "at which there should be a presumption—to be overridden only in exceptional circumstances—against additions at the institute level." The report also suggested that the NIH director "have greater budgetary authority and discretion," in order to recognize promising areas of medical research and respond to public health emergencies. In particular, the report recommended that the director be given limited authority to transfer 0.5 percent of the NIH budget across institute lines.[50]

A more controversial recommendation concerned the matter of coordinating health research at the departmental level of HHS. The Ebert group recommended that there be a Health Science Board to oversee "the health research organization, mission, priorities, and institutional management of the several elements of the Public Health Service." The committee proposed that this board should contain six members who would be appointed by the Secretary of HHS from a slate nominated by the Assistant Secretary for Health after consultation with the National Academy of Sciences and the National Academy of Public Administration. In this way, the committee hoped to insulate the proposed board from conventional political pressures. Even so, the recommendation aroused considerable controversy within the IOM Council. Some IOM members felt that any such a board would inevitably become politicized and questioned whether the issues that went into the recommendation had been "fully analyzed and developed." The Council took the unusual step of passing a formal motion to "express considerable concern about the creation of a Health Science Board and the potential harm this could cause."[51]

To be sure, the IOM opposed the politicization of NIH, whether the administration in power was Democratic or Republican. Still, there was a special sensitivity to this issue during the Reagan administration, which alone among the administrations that preceded it, appeared to hold no special brief for NIH. Simply put, the IOM had no desire to increase the Reagan administration's oversight of the National Institutes of Health. Instead, IOM members hoped to retain the system of support for medical research embodied in NIH, with its use of such devices as peer review to award grants and its insistence on the predominance of professional, disciplinary politics over conventional, electoral politics. The question that preoccupied the IOM was how best to preserve NIH, not how to change it. In the end, the people who worked in Secretary Heckler's office could make common cause with the IOM over the need to protect NIH from the

congressional desire to expand the number of institutes according to
the political whims of the moment, yet even this collaboration
contained its share of tension.

In time, the Institute of Medicine learned how to highlight, rather
than hide, its differences with the Reagan administration. To do so,
the IOM crafted its message in a way that respected the public's bias
against large government expenditures and showed an appreciation of
the media's importance in publicizing the message. Perhaps the most
visible and successful project of this type concerned the subject of low
birthweight. In undertaking this project in the winter of 1983, the
IOM understood that babies were popular objects of public sympathy.
People of all political persuasions agreed that children deserved a
healthy start, and those who followed issues in public health knew
that the rates of infant mortality and morbidity were higher in the
United States than in other developed countries. It therefore became
possible for the IOM to raise funds for this project from private
sources, such as the Commonwealth Fund, the Ford Foundation, and
the March of Dimes, and from public sources such as the National
Institute of Child Health and Human Development. Richard E.
Behrman, dean of Case Western Reserve Medical School, agreed to
chair the panel that produced a report by the spring of 1985.[52]

The report, *Preventing Low Birthweight,* stressed that preventive
measures to decrease the incidence of low birthweight babies were
both cost-effective and a means of reducing infant mortality. Although
the infant mortality rate had been cut almost in half between 1965
and 1980, much of the improvement had come about through
advances in neonatal intensive care. Neonatal intensive care units,
populated mainly by premature babies, cost a great deal for hospitals
to run and communities to support. The IOM committee suggested
that a new emphasis be placed on prevention before the fact rather
than on treatment after the fact. Such a strategy "may well prove to
be considerably less costly, both socially and economically, than
additional [expenditure] in neonatal care." Prevention, in other words,
was a public health strategy that had the characteristics of an
investment: it promised to pay future dividends, and failure to
undertake this investment would cost the nation money. The IOM
group urged, therefore, that its recommendations not be considered
pleas for additional government expenditures, but rather a means of
reducing future government expenditures. In this manner, the IOM
came to understand how to make a case for public health in a
conservative era.

The IOM report contained many specific suggestions on ways to
prevent low birthweight. People at high risk, such as teenagers,

should receive access to health education and family planning services so as to prevent unwanted pregnancies. All pregnant mothers should be given regular prenatal care, if necessary by making changes in Medicaid reimbursement policies in particular states. The committee reported that the "overwhelming weight of the evidence is that prenatal care reduces the risk of low birthweight," a finding strong enough "to support a broad, national commitment to ensuring that all pregnant women in the United States, especially those at medical or socioeconomic risk, receive high-quality prenatal care." All pregnant mothers, according to the committee, should also take prudent steps to reduce the risk of bearing a low birthweight child by not smoking or drinking during their pregnancy. To make people aware of the risks, the nation should launch a "long term highly visible public information program." For each of these measures that it recommended, the committee attempted to show that the benefits outweighed the costs.[53]

What distinguished the low-birthweight study from other IOM projects was that the organization did not let the process stop with issuing a report. Instead, it took aggressive efforts to disseminate the report. Securing additional funds, the IOM staff, headed by Sarah Brown and Enriqueta Bond of the Division of Health Promotion and Disease Prevention, prepared a short glossy summary that could be distributed easily. Although such a practice eventually became routine for IOM reports that promised to be of wide interest, it began with the low-birthweight study, a project in which Fred Robbins took a deep personal interest. The IOM also chose to release the report in dramatic fashion. Instead of simply issuing a press release or holding a press conference, the IOM delivered its findings at a hearing conducted by Representative Henry Waxman's (D-Calif.) Subcommittee on Health and the Environment of the Committee on Energy and Commerce. Held at the National Children's Medical Center on the outskirts of Washington, D.C., the hearing, staged to dramatize the problem of low-birthweight, featured the presentation of low-birthweight infants from the hospital's neonatal intensive care unit. The sight of the babies in their incubators made for a particularly touching scene. Coupled with advanced notice of the report's contents to the press and aided by Sarah Brown's skill at public relations, the dissemination strategy yielded articles in the *New York Times* and *Washington Post* and segments on the "CBS Evening News" and the "McNeil–Lehrer" Report. Follow-up reports appeared on the "CBS Morning News" and in three separate editorials in the *New York Times*.[54]

Although the IOM welcomed this publicity, Brown and Bond were dismayed by the way Representative Waxman used the report for partisan purposes. When they walked out of the Waxman hearing, they realized that Secretary Heckler, who was scheduled to receive an IOM briefing on the report that afternoon, would be displeased. During the briefing, the Secretary's staff did not show a lot of interest in the report; one senior staff member attended to his paperwork during the IOM presentation. After listening politely, Secretary Heckler asked Fred Robbins to remain behind at the end of the meeting. When the two were alone, she proceeded to admonish Robbins for testifying before Waxman's committee before she had been briefed. The Secretary did not want to give a publicity advantage to the Democrats.[55] Indeed, many of the report's recommendations rubbed against the grain of the Reagan administration. Although no one could oppose the goal of preventing low birthweight, the expansion of Medicaid remained a controversial matter. The administration sought to cap Medicaid spending and possibly to "federalize" the program and remove authority from the states; the IOM proposed to raise the level of Medicaid coverage and, in all likelihood, to increase its costs. Representative Waxman's staff later cited the report as the major supporting documentation for the Medicaid Infant Mortality Reduction Amendments of 1985 that the congressman introduced.

Although the IOM might previously have hesitated to engage in as self-conscious a public relations effort as the one that followed the release of the low-birthweight report, the increasing sophistication of the White House and other Washington centers of power in managing the news convinced the IOM that in order to be heard, the message had to be scripted in ways the media could readily interpret. Indeed, the IOM staff had wanted to engage in such activities even before the Reagan presidency. This meant photo opportunities and clear lines of argument that were not undercut by qualifications. In the past, the IOM had stressed the ambiguities and lacunae in the data for fear of placing advocacy above science. The low-birthweight study showed a new willingness to craft a message for media consumption. In so doing, the IOM engaged the Reagan administration on its own terms in order to make a case for public health entitlements, not as dead weights on the economy, which the Reagan administration believed them to be, but as sources of economic opportunity.

Congress remained an important audience for the IOM's endeavors, as the study on the regulation of nursing homes demonstrated. Passage of Medicare in 1965 had put the federal government in the business of regulating nursing homes. Before a facility could receive federal funds, it had to receive federal accreditation. Every year, state health officials, under contract to the federal government, inspected nursing homes to determine whether they met minimum health and safety standards. In May 1982, the Health Care Financing Administration announced that it wished to change some of the regulations, for example, by reducing the frequency of inspections and substituting accreditation by a private group for federal accreditation. Democratic members of the House responded by introducing a bill in the spring of 1983 that called for the IOM to study nursing home accreditation. Although the bill failed to pass, HCFA agreed to sponsor the IOM study. In the fall of 1983, HCFA and IOM signed a contract, and the study was launched. With a budget of more than $1.5 million and a staff of eight headed by David Tilson of the Division of Health Care Services, the project became a major IOM undertaking in the years between 1983 and 1986.[56]

The steering committee, headed by Sidney Katz, director of the Institute for Gerontology Research at Brown, met for the first time in December 1983. It heard from a member of Representative Waxman's staff who emphasized that no changes in the regulations would be made until the IOM study was completed. The congressman wanted "an objective nonpartisan basis for changing nursing home regulations" from the IOM. John Rother, who worked for Senator John Heinz (R-Pa.) and later became chief lobbyist for the American Association of Retired Persons, made the argument that existing regulations focused "on all the wrong things." He hoped that the IOM would develop a process that looked at quality from the patient's point of view.[57] The session made it clear that members of Congress, not administration officials, would be the primary audience for the study.

The IOM steering committee worked its way through the complicated logistics of nursing home regulation. Complex to begin with, the process acquired even more complexity because each state administered the Medicaid program in its own way and used its own system to certify nursing homes for Medicaid participation. In addition, representatives of nursing home operators and patient rights' advocates sought to lobby the committee. To accommodate the demand to present evidence, the committee held public meetings in five cities across the country in September 1984. In the beginning of 1985, when the steering committee was finally ready to consider specific recommendations, it decided to highlight the fact that there was both a quality of life and a quality of care problem in nursing homes. Combating these problems, according to the committee, required regulation; "free market competition is insufficient in and of itself." Furthermore, the federal government should play a stronger role in the certification process to overcome what the committee described as the "federal–state–local fragmentation." The committee also decided that both the quality of life in nursing homes and the current regulatory system could be improved "without necessarily increasing costs or the supply of nursing home beds." Agreeing on these basic propositions, the committee bogged down on the specifics. All of its members believed that HCFA should revise the elements on its survey report forms to emphasize actual performance and patient outcomes rather than capacity and "paper compliance." Fewer concurred in the more expansive declaration that nursing home residents had "the right to access to the community, to go out into the community and to have visitors come into the facility. The resident has the right to transportation and to escort services."[58]

When the final report was issued in 1986, Sidney Katz received an invitation to testify before a joint hearing of the three House subcommittees with jurisdiction over the matter. Representative John Dingell (D-Mich.), the chairman of the House Committee on Energy and Commerce, said, "Thanks to the Institute of Medicine, the goal of nursing home reform that has eluded us for over 10 years seems attainable." In October 1986, the Department of Health and Human Services issued draft regulations that were based in large part on the IOM report.[59] The IOM's experience with the nursing home study showed that it was possible to influence public policy on a controversial topic at a time when the very idea of federal regulation was under heavy attack.

Fund-Raising

Despite the exhilaration of the low-birthweight and nursing home studies, the constant need to raise funds drained the energy of Fred Robbins and his staff and led to many frustrations. "As you know," Fred Robbins wrote Robert Derzon toward the end of 1984, "my greatest disappointment is that we have not made more effective progress in our fund-raising." Although Robbins, like his predecessors, devoted the bulk of his time to fund-raising, the results were discouraging. A statement of the IOM's income for programs and projects showed a small nominal increase between fiscal years 1980 and 1985. Adjusted for inflation, the amount that the IOM received declined in these years and never surpassed $2.6 million (actual dollars) in any year. Furthermore, government, rather than private funds, predominated. In calendar year 1984, only 26 percent of the

IOM's income came from private sources; the largest single contributor in that year was the National Institutes of Health. Although this sort of government money and money from the NAS overhead pool kept the IOM going, it came at the cost of not allowing the IOM to do the sorts of projects its members wanted. Unrestricted grants were hard to come by, despite the continued generosity of the Robert Wood Johnson, Kellogg, and Richard King Mellon foundations and the Commonwealth Fund.[60]

Robbins did manage to bring more foundations into the IOM stable. His greatest successes came with the MacArthur Foundation, which gave the IOM $225,000 in unrestricted money at the end of 1984, and the Pew Foundation. Even before Robbins became head of the IOM, he received an invitation from the Pew Memorial Trust to advise the foundation as it considered a new initiative to "improve the health of the people of this country." In time, the foundation, which had previously been tied to the Philadelphia area and very conservative in its outlook, broadened its horizons and professed an interest in "developing and looking at innovative directions in the general field of health." If the foundation wanted fresh ideas, the IOM was more than willing to supply them. Skillful cultivation of the Pew Foundation eventually paid off in the form of a grant for $383,000 and support for the IOM's study of nonprofit enterprise in health care.[61]

Such grants helped, but not enough. The IOM turned to other sources of money, such as charging IOM members "user" fees for the services they received. David Hamburg had been reluctant to charge dues, arguing that because members devoted so much of their time to the Institute, they should not also have to pay for membership. If they wished to make voluntary contributions, they could. In response, the IOM actually dropped some of its fees in the Hamburg era, for example, no longer charging members to attend the annual reception. By way of contrast, Fred Robbins felt that, as an IOM committee phrased it, the "time is past due to ask more from members in the form of financial support." At the end of 1984, the IOM Council decided to charge annual dues and institute a registration fee for the annual meetings.[62]

User fees made only the slightest of dents in the IOM's chronic financial difficulties. Each year, the organization struggled to balance its budget by the end of the fiscal year. In fiscal year 1982, the organization averted a deficit by making some reductions in staff and spending multiyear foundation grants sooner than anticipated. At the beginning of 1983, Charles Miller, whom Robbins had named as executive officer in place of Karl Yordy, outlined the budget for the next fiscal year and announced that there could be no growth in the

IOM's core staff. Even the chiefs of the four divisions would have half of their salaries paid from project funds, meaning that they would have to devote half of their time to specific funded projects, rather than to more creative activities that might help the IOM move in new directions. The next year produced a similar sort of crisis. Lincoln Moses, chairman of the IOM Budget Subcommittee, said that unless additional revenues were raised, a deficit of $600,000 would occur in the fiscal year 1985 operating budget. The problem was that the multiyear unrestricted grants from foundations were coming to an end, and the foundations, insisting that the IOM raise its own endowment, refused to make new ones. As Moses noted, it was impossible to raise an endowment in time to save the 1985 budget. By cutting back on expenses and forcing division chiefs to spend even more of their time on specific projects, the IOM barely managed to balance the budget for that year. These annual skirmishes could be stopped only by a sizable infusion of money. No wonder, then, that Robert Derzon called funding the "most serious issue facing IOM" in 1984.[63]

The Sproull Report

By the end of 1983, the consortium of foundations that supported the IOM appeared to be getting restless, dissatisfied with the IOM's progress. In January of that year, Robert Ebert, former dean of the Harvard Medical School and current president of the Milbank Memorial Fund, wrote to Frank Press, president of the National Academy of Sciences, and proposed a study of the Institute of Medicine. It would be run by Press and the NAS, with financial support from the eight foundations that had an interest in the IOM. The idea was to conduct the study during 1984, simultaneously with the effort to find a successor to Fred Robbins. Already the age of a senior member, Robbins had no desire to seek a second term, and in May 1984, he put Theodore Cooper in charge of a search committee for a new president. With Robbins leaving, it appeared a good moment for Press to put his stamp on the IOM or, as some people feared, to abolish it altogether. Press, it seemed, was no more a fan of the IOM than Handler had been. For these reasons, Press became an active collaborator in stimulating the request for a study of the IOM and shaping its terms. Ebert let Press edit his January letter, in which Ebert wrote that the foundations would like to learn more about the definition of the IOM's mission, the qualifications of IOM members, the effectiveness of the IOM organization and staff, and the adequacy of the IOM's finances to fulfill its mission. Ebert and Press both saw the study as a high-visibility endeavor, with an eminent figure such as Clark Kerr, Elliott Richardson, Cyrus Vance, Abraham Ribicoff, or McGeorge Bundy as chair. Although Fred Robbins and his staff worried behind the scenes about the outcome, Robbins also appeared to be a supporter of such a study. "I'm delighted you are responsible for this," he wrote Ebert, who had been his colleague as a medical school dean.[64]

The project gained bureaucratic momentum in the winter and spring of 1984. Both the IOM and the NAS councils approved the project, and the foundations, such as the Commonwealth Fund, made small grants to fund it. The IOM Council, which could hardly refuse such a request from the very foundations that gave money to the IOM, thought that the study would be "valuable" and authorized its own simultaneous review of the Institute's mission, structure, and functions. Robbins optimistically told IOM members that the study would be an "important opportunity for the IOM" that would "reaffirm the value of this very special institution" and "provide a most powerful platform for the Institute's case for endowment gifts from prospective donors." As Robbins and the staff realized, however, investigations of the IOM by the NAS had rarely turned out well because of the Academy's desire to have the IOM conform to its image, which meant replicating the pattern of an honorific academy and a separate research arm.

Although the NAS could not get the eminences that Ebert had mentioned to chair the study, it did manage to recruit Robert Sproull, a distinguished physicist at Cornell and president of the University of Rochester. Among others on the nine-person committee were Robert Ebert; NAS Vice President James Ebert; William Hubbard, president of the Upjohn Company; Daniel Nathans, a Nobel Prize-winning molecular biologist from Johns Hopkins; Don Price, a professor and former dean of Harvard's Kennedy School of Government; and Elmer Staats, a distinguished public administrator and former Comptroller General of the United States. The group resembled a visiting committee that might be asked to examine a program or department at an Ivy League university.[65]

The committee soon acquired its own staff, including MIT political scientist Harvey Sapolski, and began to hold what amounted to hearings on the IOM. To prepare for these hearings, the IOM devoted a Council meeting in May to a discussion of its future direction. The session provided a very candid review of the organization's strengths and weaknesses, one similar in tone to the retreats that both David Hamburg and Fred Robbins had held at the beginning of their presidencies. The familiar litany of problems was cited. Unless the IOM acquired an endowment, it would not be able to continue in its present form. Funding was the first priority and had been the IOM's biggest problem for years. The IOM president was burdened with both programmatic and financial responsibilities and needed help. IOM reports were uneven in tone and the good ones were not distributed widely enough. Getting a study approved by the National Academy of Sciences was relatively easy if it stemmed from a government request but much more difficult if it originated from an idea generated within the IOM. Staff morale suffered because there was constant pressure to find another project on which to work and little time for professional development. Private industry knew nearly nothing about the IOM, and much more could be done to interest Congress in its work; the posture of the IOM in the broad scientific community was "poor."[66]

When Fred Robbins appeared before the Sproull committee on June 5, 1984, he left most of the dirty linen at home. Once again, he told the group that he "welcomed" the study. He emphasized the point that the IOM's greatest problem was the lack of an endowment and suggested that it would take $20 million to generate an income sufficient to replace the core funds that were provided by foundations. In discussion with the members of the committee, Robbins defended the basic principles of the IOM, in particular the concept of a mixed membership that actively participated in studies. After Robbins left the room, Robert Ebert raised some troubling questions for the committee to consider. He wondered, for example, about the causes and effects of frequent presidential turnover at the IOM and about the quality of the IOM staff. Frank Press raised the question of whether the IOM, by mixing social and scientific concerns, could be accused of having a policy bias. To study these and similar questions, committee members agreed to read a small sample of IOM reports and to have Sproull and his staff interview about 40 people, influential in health policy and the health sciences, to gain their impressions of the IOM.[67]

Perhaps the most important of these interviews was done by Sproull himself and included key officials from the Robert Wood Johnson Foundation and the Commonwealth Fund. These people spoke, in effect, as the proprietors of the IOM, and it was clear that they were far from pleased. David Rogers, president of the Robert Wood Johnson Foundation, said he was very concerned about the IOM's "second-class staff," which combined with a busy membership to form a "faulted" method of operation. Rogers said that the IOM would be better off if it adopted the model he followed at his foundation. There should be a small staff that shipped much of the work to outside organizations—a "smart" staff, similar to the one at the National Bureau of Economic Research, that was capable of publishing in respected journals. Rogers added that the IOM should be famous for the studies that it initiated. As matters stood, according to Thomas Maloney of the Commonwealth Fund, the IOM produced "second-class analysis of facts" and gave unwanted advice to the government. Maloney said he could not imagine the IOM ever recommending less health care or facing up to the trade-offs between dollars for medicine and funds for other social initiatives. In a word, the IOM was "predictable." In his summary of the conversation, Sproull noted David Rogers's remark that he would hate to see the IOM disappear, but Sproull was left with the impression that the "combination of the Reagan administration and the disenchantment" of the foundations would make it impossible for the IOM to continue with its present mode of operations.[68]

A subsequent conversation with John Sawyer, president of the Andrew Mellon Foundation, confirmed Sproull's impression. According to Sproull, Sawyer was "weary" of the IOM. Before the IOM could advance, Sawyer thought it would have to resolve its questions of leadership and be willing to make courageous studies of large issues. At the same time, Sawyer, like Rogers, implied that he would like to support the IOM, even hinting that the Andrew Mellon Foundation might give the IOM a million dollars if "we saw a plan and a leadership that would address the major issues."[69] Alvin Tarlov, head of the Kaiser Family Foundation, said that his foundation also would like to give money to the IOM. "An aggressive, optimistic, invigorating president" could get a million dollars from Kaiser.[70]

Staff member Harvey Sapolsky, who had been a colleague of Frank Press's at MIT, did the bulk of the interviews for the Sproull committee. Taken together, they constituted a damning indictment of the IOM in which even those with a close relationship to it outdid themselves in their candor. Stuart Altman said that the IOM was a "marginal force at best" in health care policy and that it played no role in financing issues. John Ball of the American College of Physicians noted that IOM committees could give you a consensus view of a problem but were "never bold." Larry Brown, a Brookings staff member, called the impact of the IOM "generally quite weak." E. Langdon Burwell, the Cape Cod practicing physician and IOM member who had often been critical of the Institute, told staff member David Caulkins that the IOM was "dominated by individuals from the academic community who speak their own language on health affairs," had "an anti-AMA bias," and produced "predictably bland" studies. John Cooper, head of the Association of American Medical Colleges, described the IOM as the "UN of health policy"—a "babble of voices." Loretta Ford, dean of nursing at the University of Rochester, complained that the IOM will do "anything they can get the money for." Economist Victor Fuchs told Sapolsky that the IOM "hasn't had the guts to call things as they are." William Gorham, head of the Urban Institute, said that the quest for funds dragged the organization "into topics where it lacks expertise and advantage." Walter McNerney called the IOM "an old-boys' club resistant to change and with lackluster leadership."[71]

As Sapolsky discovered, both the relatively liberal founders of the IOM and the relatively conservative officials of the Reagan administration found fault with the IOM. Rashi Fein of the Brookings Institution lamented the fact that the IOM took establishment positions, whether liberal or conservative, and avoided the big issues such as health care financing and access to health care. James Shannon complained that the Institute of Medicine failed to provide the "informed protection" that the NIH needed. "It's OK to be concerned about social issues," said Shannon, in a variation of a speech he had been giving since 1969, but the IOM was "dominated by these concerns to the detriment of medical science issues." Reagan administration officials, such as Edward Brandt, the Assistant Secretary of Health who described himself as a strong IOM supporter, told the Sproull committee in direct testimony that the IOM should do its work "quicker and cheaper." Roger Bulger conceded that the "IOM had been quite close to the Carter administration." Robert Rubin, a medical doctor with an academic background who had served as Assistant Secretary of Planning and Evaluation in the Department of Health and Human Services during the Reagan administration, described the IOM's studies as "not politically balanced." For this reason and because of its slow pace and undistinguished staff, Rubin believed, the IOM's influence had declined during the Reagan administration. Karl Yordy admitted that the Reagan administration looked at the IOM as the home of failed Democrats.[72]

Not everything that Sproull and his staff heard was negative. John Ball of the American College of Physicians, although quite critical, nonetheless admitted that the IOM provided a "balanced, independent view." The Brookings Institution represented the Democrats in exile, and the American Enterprise Institute was the home of the Republicans in exile. The "Beltway bandits," a common epithet for the independent consulting firms surrounding Washington, only "give you what you want." By way of contrast, the Institute of Medicine was far more independent in its operation and outlook. Enriqueta Bond of the IOM staff told Harvey Sapolsky that although the IOM was hardly a fast-response organization, it "could handle the hard topics." Robert Derzon noted that the IOM's troubles were "financial and managerial" not structural. Stuart Bondurant of the University of North Carolina Medical School said that many of the IOM reports had been "solid and several have clearly influenced health policy on important issues." Vincent Dole, senior professor emeritus at Rockefeller University, argued that the "IOM or some other leadership group with similar objectives is indispensable." Still, almost no one, in either the interviews, the direct testimony, or the many letters that the Sproull committee received, offered unqualified praise. The lingering impression was of a troubled and unfocused organization. "It is not clear to me," summed up Marsh Tenney of the Dartmouth Medical School, "that IOM serves any useful purpose. If it were to disappear, I doubt if anyone would note a change."[73]

By the end of the summer, IOM staff members had heard enough about the Sproull committee's deliberations to know that the organization faced a real crisis. Karl Yordy told Charles Miller that when the IOM Council met with Robert Sproull in September, it would be surprised by his "preliminary negative conclusions." Yordy argued that the Council should be prepared to alter these conclusions. Sure enough, Sproull laid out many criticisms of the IOM during the September meeting. Speaking bluntly, he told the Council that some of the IOM reports were not very good, that the IOM had not met the goals of its founders, and that the foundations were frustrated with the IOM. Furthermore, the IOM's fund-raising efforts had been hampered by the "parade of presidents" and by the IOM's own "amateurish" actions. Hinting that it might be better if the IOM became a "strictly honorific society," he concluded that it was unlikely the IOM would "survive as a major national force unless major changes are made."[74]

The IOM did its best to respond to this onslaught and to soften what promised to be a very critical report. Council members drafted a letter in which they signaled their desire to work on a restatement of the IOM mission; to engage in a "rigorous, hard-nosed review of the budget and staff"; to increase the scientific and technical emphasis in the IOM; and to work harder on disseminating the results of IOM studies. Robbins followed this letter with another that listed the IOM's latest achievements, such as entering into a cooperative agreement with the Health Care Financing Administration to support the four advisory boards and obtaining funds for the dissemination of the low-birthweight study. In the meantime, IOM task forces on mission and organization and on fund-raising and budget met to prepare a quick response to the Sproull report when it appeared in November.[75]

A short and relatively informal document, the Sproull report, or more properly the "Final Report of the Study Committee on the Institute of Medicine," contained recommendations that, simply put, were devastating to the IOM. "In general," the committee wrote, "we propose a strengthening of the Institute of Medicine that amounts to a rebirth." In its second life, the Institute of Medicine would be resurrected as the National Academy of Medicine. The IOM's studies would be transferred to the National Research Council and in their place would come a "policy-oriented core program, flexibly and imaginatively attacking the great health and medical problems and making better use of an even more distinguished membership." The committee referred to this core program as a "health policy center that would serve as a crucible for national and international health policy discussions." In order to move from an Institute of Medicine to a National Academy of Medicine, the committee proposed that a new emphasis be placed on "professional distinction as judged by professional peers," rather than on "the visibility and influence of a candidate's position." In its initial phase, the National Academy of Medicine would have about 150 members, chosen by the NAS Council. Those IOM members not selected for the new academy would become senior members of the National Academy of Medicine for a limited term.[76]

If the Sproull committee's recommendations were adopted, it would mean the end of the Institute of Medicine. Walsh McDermott's experiment would end in failure, and something more compatible with the other parts of the National Academy of Sciences would take its place. The basic idea of moving the IOM's studies to the National Research Council was not new. It had been proposed when the IOM was first founded and again in a different way by the Ebert report at the beginning of the 1980s. The difference with the Sproull committee was that it had the eyes and ears of the foundation community.

Defending the Institute of Medicine

Members of the IOM community reacted with concern. Karl Yordy called the report a "key event in the history of the IOM" and urged Robbins "to set a new course, not give up the ship." Using a different military metaphor, Yordy argued the need to "move rapidly to develop a defense" against the "blitzkrieg" of NAS President Frank Press before he conquered the IOM. Press wanted to disarm his opposition in the same manner that Hitler conquered Poland. "I think the IOM must become like Switzerland," Yordy asserted, "too well armed to be worth the effort for a Kaiser or a Hitler to conquer." Yordy accused Press of acting in bad faith by outmaneuvering the foundations and twisting "their original intent toward his own internal organizational objectives" and by "cynically" using the Sproull committee "to generate an occasion for the action he wanted to take all along." The IOM would now have to play Winston Churchill to Sproull's Neville Chamberlain.[77]

Although other IOM insiders did not put the matter in such graphic terms, many agreed with Karl Yordy. Alan Nelson, a Council member who specialized in internal medicine and endocrinology, took exception to the study committee's implication that "health policy is not anything that good scientists would want to mess with." William Bevan, a vice president of the MacArthur Foundation and a Council member, said he was disappointed in the report and believed it had the "capacity for great mischief." Philip Leder, head of the Department of Genetics at Harvard, reacted with "shock, surprise, and amazement at the weaknesses in concept and structure alleged by the report and the radical scope of the proposed solutions." Ruth Gross, a professor of pediatrics at Stanford found the report "difficult to evaluate" and "quite confusing," and Ben Lawton, a surgeon from the Marshfield Clinic, reported that his "initial reaction was outrage! From our briefing by Dr. Sproull, I knew we could expect considerable negative comment. I did not expect a frontal attack and proposal to abolish the Institute of Medicine and emasculate its proposed successor."[78]

If Lawton expressed the feelings of the majority, this still left room for a considerable minority. Gilbert Omenn, dean of the School of Public Health and Community Medicine at the University of Washington and head of the IOM Task Force on Mission, Structure, and Finances, found the Sproull report "very supportive of the IOM" because their "crucial finding was that the IOM (or something very much like it) is, indeed, needed." On the matter of the plan to create a National Academy of Medicine, Omenn, who had worked with Frank Press in the Carter White House, said he would keep an open mind. Agreeing with Omenn, Robert Butler, a Council member and the former head of the National Institute of Aging, thought the Sproull committee did "a good job and came up with some respectable conclusions."[79]

This set up the major lines of division for an IOM Council meeting in January 1985 devoted to formulating the IOM response to the Sproull report. In general, those who wished to defend the IOM

outnumbered those who thought the IOM should react
sympathetically to the report. The Council objected in particular to
the notion that the IOM had an incompatible mission of trying to
combine scientific objective analysis on the one hand and subjective
policy judgments on the other. In the view of many IOM members,
such distinctions perpetuated the nineteenth century notion that
science was an empirical, value-free endeavor and ignored the
realities of both the scientific and the policy processes. Rejecting this
central premise of the Sproull report, the Council went on to argue
that there was no need to move the IOM's studies to the National
Research Council. The idea of a health policy center had merit, but it
could easily be accommodated within the existing IOM structure. This
was something that the IOM hoped to explore, just as it pledged to
improve the quality of its reports and refine its criteria for
membership. The Council noted that although other things could be
done to improve the way in which the IOM functioned, substantive
changes should await the arrival of the next president. In summary
then, the IOM argued that the Sproull report was based on a
misperception of the IOM's work, and it opposed the transfer of IOM
projects to the NRC.[80]

Based on the Council's discussion and the ongoing work of its task
forces, the IOM Council composed a formal statement in response to
the Sproull report that appeared at the end of January. In this
statement, the Council tried to be deferential toward the Sproull
committee—"a distinguished group which addressed its task sincerely
and thoughtfully" and positive about the IOM—we "want to build on
the accomplishments of the past 14 years." For all of this deference,
the IOM Council felt compelled to "disagree with many of their
recommendations and some of their findings." The statement
proceeded to object to the report's basic premise, to argue against
moving the IOM's studies to the NRC, and to oppose the creation of a
National Academy of Medicine. The IOM gave ground only in the
areas of membership, in which it conceded that improvements could
be made, and finances, in which it admitted that although it always
balanced its books by the end of the year, each year was a struggle.
Despite the considerable effort to be conciliatory yet firm, the Council
still could not contain all of its members. Ronald W. Estabrook, a
biochemist from the University of Texas at Dallas and a member of
the National Academy of Sciences, felt compelled to write a dissent in
which he said that he was "not disturbed" by the prospect of
transferring the IOM's studies to the NRC and that membership in
the IOM should be based on "evidence of scholarly achievement." The

notion of a mixed membership, with practicing physicians and public policy practitioners, held little appeal for him.[81]

The NAS Council served as the final arbiter on the matter, and in this forum the IOM did quite well. Remembering the difficulties with the National Academy of Engineering, the NAS Council had little inclination to create a new academy of medicine within the NAS complex and quickly dismissed the idea. As it had at the time of the IOM's creation, the NAS Council conceded that there was a need for a "membership organization directed toward medicine and health policy"—something, in other words, like the IOM. The NAS Council also approved of many of the IOM's recent decisions, such as its willingness to examine its membership criteria closely and its decision not to elect new members in 1984 but instead to put in place a new system that emphasized achievement and not merely position. On other matters, however, such as the crucial question of whether to move the IOM's studies to the NRC, the NAS Council resolved to hold off its decision until it met in February.[82]

At this February 1985 meeting, Frederick Robbins, in a rare display of emotion, vented his feelings. He told the NAS Council that the last few months had been particularly difficult. Although he had started as an enthusiastic supporter of the process, he was dismayed to receive a "highly critical" report that contained "drastic recommendations for change" and indicated "a true lack of support for the concept of the IOM." The report put him and his organization in a "defensive posture which is not a pleasant situation to be in." Robbins did not say so, but it rankled him in particular that the report cast aspersions on the IOM staff. Of all the people in the NAS complex, Robbins, who held important IOM and NRC posts, was in the best position to compare the NRC and IOM staffs, and in his opinion, the quality of the IOM staff exceeded that of the NRC staff.[83]

Whether or not the NAS Council was moved by Robbins's presentation, it rejected the recommendations of the Sproull committee and decided not to recommend that the IOM be disbanded and its studies transferred to the NRC. Instead, it opted to take a wait-and-see approach. It issued a statement announcing that it "continued to be strongly supportive of the Institute of Medicine" but recognized the issues and problems raised by the Sproull committee. In response, it proposed to join the search for the new IOM president and, together with this new president, to "further examine" the issues. At a minimum, the IOM had won a reprieve without making a major concession. All it had to do was disband its presidential search committee and replace it with one jointly chosen by Frank Press and Fred Robbins. "As is the case with any organization under new leadership," Frank Press explained to John Sawyer of the Mellon Foundation, "the opportunity for self-analysis and charting new directions is one that should coincide with the inauguration of the next IOM president."[84]

The action shifted to the presidential search committee. On March 20, 1985, Paul Marks, president of the Sloan-Kettering Cancer Center, agreed to chair the committee. Eight others joined him, including four medical doctors, two distinguished Ph.D. scientists, a public health expert, and a hospital administrator. Five of the committee members, including the chairman, were selected by the National Academy of Sciences, giving the NAS the majority of the appointments. It was this committee's job to come up with a new president of the IOM by the time of Robbins's departure on October 1, 1985.[85]

The wait stretched through the summer and into the fall. The IOM Council meeting on September 23 proceeded as usual, the chief item of business being the usual distressing news about IOM finances. The staff predicted a quarter of a million dollar shortfall in operating expenses for fiscal year 1986, which would mean that division directors and division secretaries could not be carried on the general operating budget and would instead have to be shifted to specific projects. As the meeting wound down, Frank Press walked into the room and announced that Samuel Thier, the head of the Internal Medicine Department at Yale, would become the next president of the Institute of Medicine.[86]

Conclusion

Although IOM members could not have known it at the time, the organization had weathered the crisis. During the era of Fred Robbins, the IOM managed to survive two threats to its existence in the form of two NAS reviews of its structure and accomplishments. The Sproull report represented the most profound assault on the concept of the IOM that the organization had ever faced, calling for the rebirth of the IOM. As it worked out, the IOM was reborn, not as a National Academy of Medicine but as an Institute that would rededicate itself to speaking out on health policy issues in scientifically and socially responsible ways.

The Robbins era might be compared to the Board on Medicine period. In both instances, leaders of the organization spent the bulk of their time attempting to define the IOM's basic mission and negotiating with the National Academy of Sciences over the degree of autonomy the IOM would enjoy. Because the Board on Medicine and the Robbins-era IOM devoted so much time to these endeavors, they accomplished less in the way of completing studies than they might otherwise have. In both eras, the country stood on the cusp of social and political change. The Board on Medicine began during the Great Society and concluded its work during the presidency of Richard Nixon. Robbins's tenure started at a time when the IOM enjoyed a tight relationship with the Carter administration and ended when the Reagan administration was at the height of its powers.

It was to Fred Robbins's credit that he adjusted to changes in the external environment without succumbing to despair. He learned to work with the Reagan administration, sometimes as a collaborator and sometimes as an adversary. The results were highly acclaimed studies on such topics as the organization of the National Institutes of Health and low birthweight babies. Although the course of the Sproull committee's work disheartened him, he kept the IOM on course, like the leader of a nation during a civil war. If Fred Robbins left many problems with which his successor would have to deal, he also started many projects, such as one on the future of public health, that would come to fruition under his successor. Because of Robbins's perseverance, his successor would also deal with a foundation community that was somewhat chastened by its experience with the Sproull report and more predisposed than before to consider providing the IOM with an endowment. Of all of Robbins's achievements, however, perhaps the very fact that there was an IOM to pass on to Sam Thier constituted the greatest accomplishment of all.

Notes

1. IOM Council Meeting, Minutes, March 19, 1980, Institute of Medicine (IOM) Records, National Academy of Sciences (NAS) Archives.

2. "Summary of Discussion, IOM Program Committee Meeting, March 26, 1980," Yordy Files—Second Series, Accession 94-111, IOM Records.

3. "Meeting of the Presidential Search Committee, July 18, 1979," Search Committee Files, Accession 93-192, IOM Records.

4. "Steps Taken in the 1979 Search," Search Committee Files, Accession 93-192, IOM Records.

5. William Danforth to Philip Handler, December 21, 1979, Search Committee Files, Accession 93-192, IOM Records.

6. "Nobel Laureate Named New President of Institute," IOM Press Release, March 7, 1980, Search Committee Files, Accession 93-192, IOM Records.

7. Marjorie Sun, "Institute of Medicine Gets New President," *Science,* 210 (November 7, 1980), p. 616.

8. David Hamburg to Members, Institute of Medicine, March 3, 1980, Search Committee Files, Accession 93-192, IOM Records.

9. Annual Meeting Program, 1980, in Institute of Medicine, *1981 Annual Program Plan* (photo-offset), 1981.

10. Public Law 96-171 was passed in December 1979.

11. IOM Council Meeting, Minutes, January 16, 1980, IOM Records; William Lybrand to Robert M. White, NRC Administrator, April 15, 1980, Yordy Files, Accession 94-111, IOM Records; Karl Yordy to David Hamburg, January 22, 1980, and Draft Minutes of Meeting, July 15, 1980, both in Airline Pilot Study Files, Accession 81-065, IOM Records.

12. See materials in Airline Pilot Study Files, Accession 81-065, IOM Records; Institute of Medicine, *Airline Pilot Age, Health, and Performance,* (Washington, D.C.: National Academy of Sciences, 1981).

13. "Development of Study Plan for Two-Year Study of Nursing Education," n.d., Yordy Files, Accession 94-111, IOM Records.

14. Institute of Medicine, *Nursing and Nursing Education: Public Policies and Private Actions* (Washington, D.C.: National Academy Press, 1983), pp. 2, 7, 11–13, 16, 17, 19.

15. William A. Lybrand to Dr. John H. Moxley III, Assistant Secretary of Defense for Health Affairs, February 15, 1980, Graduate Medical Education in the Military Files, Accession 81-063A, IOM Records.

16. Institute of Medicine, *Graduate Medical Education and Military Medicine* (Washington, D.C.: National Academy of Sciences, 1981), pp. 2, 13.

17. Dr. William A. Lybrand to Paul L. Sitton, Executive Officer, NAS, June 12, 1980, Health Effects of Marijuana Files, Accession 86-064-13, IOM Records.

18. "Committee Meeting Transcript," April 15, 1981, Health Effects of Marijuana Files, 86-064-13, IOM Records.

19. Robert L. Mitzenheim to Dr. Linda Dujack, March 30, 1981; Martha Stone to Dujack, April 30, 1981; Thomas E. Campbell to Dujack, March 19, 1981; and Sarah Swingell, Celeste Texas, to Dujack, May 27, 1981, all in Health Effects of Marijuana Files, IOM Records.

20. Transcript of Committee Meeting, April 15, 1981, Health Effects of Marijuana Files, IOM Records.

21. Transcript of Committee Meeting, June 2, 1981, Health Effects of Marijuana Files, IOM Records.

22. Institute of Medicine, *Marijuana and Health* (Washington, D.C.: National Academy Press, 1981), pp. 2–3, 5.

23. *Ibid.,* p. 6.

24. Vicki Weisfeld, Acting Director, Division of Health Promotion and Disease Prevention, to Members of the Council, November 3, 1980, and Minutes of the Committee on Toxic Shock Syndrome, September 11–12, 1981, both in Toxic Shock Files, Accession 86-064-11, IOM Records.

25. Institute of Medicine, *Toxic Shock Syndrome—Assessment of Current Information and Future Research Needs* (Washington, D.C.: National Academy Press, 1982).

26. IOM Council Meeting, Minutes, March 19, 1980, IOM Records.

27. Frederick Robbins to IOM Membership, November 4, 1981, in Wallace Waterfall Materials; IOM Council Meeting, Minutes, September 16, 1981, IOM Records.

28. IOM Council Meeting, Minutes, September 17, 1980; March 18, 1981; and June 2, 1980, all in IOM Records.

29. IOM Council Meeting, Minutes, June 2, 1981, IOM Records.

30. "Interim Report of IOM Task Force on Structure," IOM Records.

31. IOM Council Meeting, Minutes, September 16, 1981, IOM Records.

32. Frederick C. Robbins to Frank Press, October 14, 1981, Yordy Files, Accession 91-051, IOM Records.

33. "A Report to Frank Press on Matters of NRC Organization and Procedures," February 18, 1982, Yordy Files, Accession 91-051, IOM Records.

34. IOM Council Meeting, Minutes, March 17, 1982, IOM Records; Frederick Robbins to Frank Press, March 23, 1982, Yordy Files, Accession 91-051, IOM Records.

35. Frank Press to the Staff of the National Research Council, March 2, 1982, and "The Organization of the National Research Council," n.d., both in Yordy Files, Accession 91-051, IOM Records; IOM Council Meeting, Minutes, July 1982, IOM Records.

36. "Report to the Council: Task Force on Membership Issues," May 5, 1983, IOM Records; IOM Council Meeting, Minutes, January 20, 1981, IOM Records.

37. IOM Council Meeting, Minutes, January 16, 1980, IOM Records; Karl Yordy to David Hamburg, September 2, 1980; Fred Robbins to Fred Abby, Health Care Financing Administration, December 4, 1980; and Yordy to Arnold S. Relman, December 23, 1980, all in Yordy Files—Second Series, Accession 94-111, IOM Records.

38. IOM Council Meeting, Minutes, November 17, 1982; January 19, 1983; and September 23, 1985; and IOM Annual Report for 1984, draft copy in IOM Council Meeting, Minutes, all in IOM Records.

39. IOM Council Meeting, Minutes, March 18, 1981; May 19, 1982; and July 20, 1983, all in IOM Records.

40. IOM Council Meeting, Minutes, January 15, 1981, and July 21, 1982, IOM Records.

41. IOM Council Meeting, Minutes, July 21, 1982, and November 18, 1981, IOM Records.

42. IOM Council Meeting, Minutes, January 19, 1983; May 18, 1983; and September 21, 1983; Draft Copy of Annual Report, 1984, and IOM Council Meeting, Minutes, March 19, 1984, and July 21, 1982, all in IOM Records.

43. IOM Council Meeting, Minutes, November 18, 1961, IOM Records; and Bradford Gray to John T. Dunlop, April 8, 1983; Fred Robbins to Kingman Brewster, March 9, 1983; "Draft Proposal to the Pew Foundation," n.d.; and Gray to Thomas F. Frist, Jr., M.D., President and CEO, Hospital

Corporation of America, December 15, 1982; March 14, 1983; and June 14, 1983, all in For-Profit Enterprise in Health Care Files, Accession 91-007-4, IOM Records.

44. "Minutes of Meeting," July 29–31, Airlie House, For-Profit Enterprise in Health Care Files, IOM Records.

45. *Ibid.*

46. Committee Meeting, Minutes, October 1–2, 1984, For-Profit Enterprise in Health Care Files, IOM Records.

47. Bradford Gray and Walter J. McNerney, "For-Profit Enterprise in Health Care: The Institute of Medicine Study," *New England Journal of Medicine* 314 (June 5, 1986), pp. 1525, 1526, 1528.

48. Form letter from Samuel Thier to financial contributors to the For-Profit Enterprise in Health Care Study, August 1, 1986, For-Profit Enterprise in Health Care Files, IOM Records.

49. Institute of Medicine, *Responding to Health Needs and Scientific Opportunity: The Organizational Structure of the National Institutes of Health* (Washington, D.C.: National Academy Press, 1984).

50. *Ibid.,* pp. 5, 10, 20.

51. *Ibid.,* p. 26; IOM Council Meeting, Minutes, September 17, 1984, IOM Records.

52. Fred Robbins to Thomas W. Moloney, Commonwealth Foundation, May 29, 1985, Low-Weight Birth Files, Accession 86-064-07, IOM Records.

53. Richard E. Behrman, Frederick C. Robbins, and Sarah S. Brown, "Preventing Low Birthweight," mimeo dated May 1985, Low-Weight Birth Files, IOM Records.

54. *Ibid.*

55. Fred Robbins to Wallace K. Waterfall, March 20, 1990, Waterfall Materials.

56. IOM Council Minutes, July 20, 1983, IOM Records; "Proposal to Study the Regulation of Nursing Homes," September, 1983, Nursing Home Regulation Files, Accession 91-007, IOM Records.

57. Steering Committee Meeting, Minutes, December 7, 1983, Nursing Home Regulation Files, IOM Records.

58. Steering Committee Meeting, Minutes, February 20–21, 1986, Nursing Home Regulation Files, IOM Records.

59. Institute of Medicine, *Annual Report, 1986, Program Plan, 1987* (Washington, D.C.: National Academy Press, 1987), p. 21.

60. Fred Robbins to Robert Derzon, July 31, 1984, Development Office Files, Accession 91-045, IOM Records; financial data from IOM Annual Report, 1984.

61. James M. Furman, Executive Vice President, MacArthur Foundation, to Fred Robbins, November 14, 1984, Development Office Files, Accession 91-045, IOM Records; Robert I. Smith, President, Glenmede Trust Company, to Robbins, August 16, 1979; Smith to Robbins, November 30, 1979; David Hamburg to Smith, July 3, 1980; Smith to Robbins, November 24, 1981; and Glenmede Trust Company to Robbins, May 7, 1985, all in Development Office Files, Accession 91-029, IOM Records.

62. "Membership Dues and Annual Meeting Registration Fees," n.d., but probably November 1984, Accession 91-051, IOM Records.

63. IOM Council Meeting, Minutes, March 19, 1982; January 19, 1983; March 9, 1983; January 16, 1985; and July 16, 1984, all in IOM Records.

64. Fred Robbins to Robert Ebert, January 19, 1984, Yordy Files, Accession 91-051, IOM Records.

65. "Minutes of NAS Staff Meeting," March 22, 1984, Waterfall Materials; Frank Press to Dr. Robert J. Buchanon, March 30, 1984, Yordy Files, Accession 91-051, IOM Records; Fred Robbins to IOM Members, March 30, 1984, Waterfall Materials; Margaret E. Mahoney to Frank Press, July 9, 1984, Sproull Committee Files, Accession 89-013-05, IOM Records.

66. IOM Council Meeting, Minutes, May 13–14, 1984, IOM Records; Charles Miller to IOM Professional Staff, May 25, 1984, Waterfall Materials.

67. Staff Summary, Committee Meeting, June 5, 1984, Sproull Committee Files, Accession 89-013-05, IOM Records.

68. R. L. Sproull, "Interview with David Rogers, Robert Blendon, Margaret Mahoney, and Thomas Maloney," July 10, 1984, Sproull Committee Files, IOM Records.

69. Robert L. Sproull, Confidential Memorandum of a Conversation with Dr. John Sawyer, President, Andrew Mellon Foundation, August 9, 1984, Sproull Committee Files, IOM Records.

70. Harvey Sapolsky, Interview with Alvin Tarlov, August 22, 1984, Sproull Committee Files, IOM Records.

71. Harvey Sapolsky, Interview with Stuart Altman, July 27, 1984; Sapolsky, Interview with John Ball, July 11, 1984; Sapolsky, Interview with Larry Brown, n.d.; Sapolsky, Interview with E. Langdon Burwell, July 15, 1984; Sapolsky, Interview with John Cooper, July 11, 1984; Robert Sproull, Interview with Loretta Ford, n.d.; Sapolsky, Interview with Victor Fuchs, September 7, 1984; Sapolsky, Interview with William Gorham, July 20, 1984; and David Calkins, M.D., Interview with Walter McNerney, October 30, 1984, all in Sproull Committee Files, IOM Records.

72. Harvey Sapolsky, Interview with Rashi Fein, June 29, 1984; Sapolsky, Interview with James Shannon, September 6, 1984; Minutes of Committee Meeting, June 16, 1984, in David Calkins to Study Committee on the Institute of Medicine; Larry McCray, Interview with Robert Rubin, July 2; 1984; and Sapolsky, Interview with Karl Yordy, July 25, 1984, all in Sproull Committee Files, IOM Records.

73. Harvey Sapolsky, Interview with John Ball, July 11, 1984; Sapolsky, Interview with Enriqueta Bond, July 12, 1984; Sapolsky, Interview with Robert Derzon, September 6, 1984; Stuart Bondurant to Robert Sproull, August 7, 1984; Vincent Dole to Sproull, August 31, 1984; and S. Marsh Tenney to Sproull, August 14, 1984, all in Sproull Committee Files, IOM Records.

74. Karl Yordy to Charles Miller, August 23, 1984, Yordy Files, Accession 91-051, IOM Records; Miller to Members of the IOM Council, August 28, 1984, Waterfall Materials; Meeting of the IOM Council, Executive Session, September 16–17, 1984, IOM Records.

75. Fred Robbins to Robert Sproull, October 3, 1984, Waterfall Materials; IOM Council Meeting, Minutes, November 19, 1984, IOM Records.

76. "Final Report of the Study Committee on the Institute of Medicine," National Academy of Sciences, Washington, D.C., November 1984, pp. 3, 9.

77. Karl Yordy to Fred Robbins, December 20, 1984, and Yordy to Robbins, February 4, 1985, Yordy Files, Accession 91-051, IOM Records.

78. William Bevan, "Comments on Sproull Report," January 13, 1985; "Memorandum to File, Conversation with Dr. Phil Leder," January 2, 1985; Frederic Solomon to Charles Miller, January 3, 1985; "Transcribed Phone Call, Ben Lawton's Reaction to Sproull Committee Report," January 7, 1985; and Fred Solomon to Charles Miller, January 2, 1985, all in Yordy Files, Accession 91-051, IOM Records.

79. Gil Omenn to Fred Robbins, December 19, 1984, and "Comments Dictated 1/3/85 by Bob Butler," both in Yordy Files, Accession 91-051, IOM Records.

80. IOM Council Meeting, Minutes, January 13–14, 1985, IOM Records.

81. "Statement of the Council of the Institute of Medicine to the Council of the National Academy of Sciences," January 29, 1985, and Ronald W. Estabrook to Members of the National Academy of Sciences, January 25, 1985, both in Waterfall Materials.

82. Philip M. Smith to Members of the Council, National Academy of Sciences, January 30, 1985, Waterfall Materials.

83. Frederick Robbins, "Statement Before the Council of the National Academy of Sciences, " February 8, 1985, Waterfall Materials.

84. Fred Robbins to Members of the Institute of Medicine, February 11, 1985, and Frank Press to John Sawyer, February 11, 1985, both in Waterfall Materials.

85. Frank Press and Fred Robbins to Paul A. Marks, March 20, 1985, Waterfall Materials.

86. Minutes of IOM Council Meeting, July 15, 1985, and September 23, 1985, IOM Records.

Irvine H. Page's group to discuss the formation of a National Academy of Medicine met for the first time on January 17, 1967, at the Cleveland Clinic Foundation. Although Page never realized his ambition to create such an academy, this group supplied the impetus that eventually led to the founding of the Institute of Medicine. Top, left to right: Fay H. Lefevre, M.D.; J. Englebert Dunphy, M.D.; Carleton B. Chapman, M.D.; Francis D. Moore, M.D.; William B. Bean, M.D.; John B. Hickam, M.D.; E. Cowles Andrus, M.D.; Robert A. Aldrich, M.D.; Ivan L. Bennett, Jr., M.D.; and Stuart M. Sessoms, M.D. Bottom, left to right: James A. Shannon, M.D.; Frederick C. Robbins, M.D. (who would later become the Institute's fourth president, in 1980); Irving S. Wright, M.D.; Irvine H. Page, M.D.; Douglas D. Bond, M.D.; and Robert H. Williams. Photograph courtesy of The Cleveland Clinic Foundation.

Frederick Seitz, Ph.D., a distinguished physicist, created the Board on Medicine in 1967 during his tenure as president of the National Academy of Sciences (1962–1969). Photograph by Harris and Ewing, Washington, D.C., courtesy of the National Academy of Sciences.

Walsh McDermott, M.D., a professor of medicine and public health at Cornell Medical School, chaired the Board on Medicine and played a leading role in the creation of the Institute of Medicine. Photograph courtesy of The New York Hospital–Cornell Medical Center Archives.

Irving M. London, M.D., who was chairman of the Department of Medicine at Albert Einstein College of Medicine in 1967, was a key ally of Walsh McDermott on the Board on Medicine. Photograph courtesy of Irving London.

Julius H. Comroe, Jr., M.D., director of the Cardiovascular Research Institute of the San Francisco Medical Center, tended to side with Irvine Page in the debates over the creation of the Institute of Medicine. Photograph courtesy of the University of California at San Francisco.

James A. Shannon, M.D., director of the National Institutes of Health from 1955 to 1968, believed that the primary purpose of the Institute of Medicine should be to support the government's role in medical research. Photograph by Ralph Fernandez, courtesy of the History of Medicine Division, National Library of Medicine.

Robert J. Glaser, M.D., who was dean of Stanford's School of Medicine in 1967, wrote the Board on Medicine's heart transplant statement and chaired the Institute's Initial Membership Committee. He also served as acting president of the Institute until John Hogness took office in the spring of 1971. Photograph courtesy of Robert Glaser.

Irvine H. Page, M.D., who was head of the Research Division at the Cleveland Clinic from 1945 to 1967 and a leading expert on hypertension and heart disease, was a strong advocate for the creation of a National Academy of Medicine. He and Walsh McDermott engaged in spirited debates within the Board on Medicine. Photograph courtesy of The Cleveland Clinic Foundation.

Philip Handler, Ph.D., a well-known biomedical researcher at Duke University who was president of the National Academy of Sciences from 1969 to 1981, played a key role in the creation of the Institute of Medicine and in its early development. His constant goal was to maintain the integrity of the Academy. Photograph courtesy of the National Academy of Sciences Archives.

Presidents of the Institute of Medicine

John R. Hogness, M.D. (1970–1974), was the Institute of Medicine's first president. He started the IOM Council and initiated many of the routines that governed the Institute's development.

Donald S. Fredrickson, M.D. (1974–1975), a distinguished biomedical researcher, stayed only a short time at the Institute before leaving to become director of the National Institutes of Health.

Presidents of the Institute of Medicine

David A. Hamburg, M.D. (1975–1980), helped to bring the Institute to national prominence. He established the divisions that still underlie the basic organizational scheme of IOM.

Frederick C. Robbins, M.D. (1980–1985), a Nobel laureate, brought his knowledge of medical research to the Institute, helping the organization to overcome the challenge posed by the Sproull report.

Samuel O. Thier, M.D. (1985–1991), energized the Institute and led a successful fund-raising effort. As a result of his work, IOM reached parity with the other components of the Academy complex.

Kenneth I. Shine, M.D. (1992–present), ushered the Institute into its second quarter century. Now in his second term, he is IOM's longest-serving president.

Karl Yordy (left), a durable and valuable member of the Institute's staff, served the IOM from the era of John Hogness into the era of Ken Shine, both as executive officer after Roger Bulger and as director of the Division of Health Care Services. Charles Miller (right) succeeded Yordy as executive officer, serving from 1983 to 1988—years that encompassed threats associated both with possible insolvency and the Sproull report, and then increasing financial security and expansion of the Institute's program. Photograph courtesy of Jana Surdi.

The Institute's first executive officer, Roger J. Bulger, M.D., was a close associate of John Hogness. He guided the daily operations of the Institute in its early years and went on to a distinguished career in medical administration, most recently as president of the Association of Academic Health Centers. Photograph courtesy of the Association of Academic Health Centers.

Enriqueta C. Bond, Ph.D. Samuel Thier relied heavily on Queta Bond to manage the day-to-day affairs of the Institute. She served in different capacities under a number of IOM presidents before assuming the presidency of the Burroughs Wellcome Fund in 1994. Photograph courtesy of the Burroughs Wellcome Fund.

Karen Hein, M.D., a former Robert Wood Johnson Health Policy Fellow, served as the Institute's executive officer from 1995 until 1998 before leaving to become president of the William T. Grant Foundation. Photograph courtesy of Karen Hein.

Susanne Stoiber, M.S., M.P.A., was named IOM executive officer in 1998. She came to the Institute from the Department of Health and Human Services, where she was deputy assistant secretary for planning and evaluation. In a previous stint at the Academy, she directed the Division of Social and Economic Studies. Photograph courtesy of Susanne Stoiber.

An M.D.-Ph.D. with a background in cardiovascular surgery and long a major figure at the Institute, Theodore Cooper chaired the study that produced *Confronting AIDS: Update 1988.* Photograph courtesy of Pharmacia & Upjohn.

Robert A. Derzon, M.B.A., a distinguished hospital administrator, followed his service as the first head of the Health Care Financing Administration with a period as a scholar in residence at the Institute. He was an IOM Council member during Fred Robbins's tenure and helped recruit Samuel Thier. Photograph courtesy of Robert Derzon.

Robert M. Ball, M.A., head of the Social Security Administration from 1962 to 1973, became a scholar in residence at the Institute and served as a confidante of IOM presidents from Hogness through Robbins.

"Watergate"—a national preoccupation from 1972 to 1974, and a word that became synonymous with political scandal—originated with a burglary that took place on the night of June 17, 1972, in this fashionable Washington, D.C., office complex. In 1974, the Institute moved into the 6th-floor offices that the Democratic National Committee had recently vacated. Its offices attracted tourists and others interested in the "stuff of history." Photograph courtesy of Elena Nightingale.

In one of the more unusual circumstances to affect the Institute, before David Hamburg could accept the presidency of IOM in 1975, he needed to negotiate the release of some of his Stanford students who had been kidnapped by rebels in Zaire. He is shown here back in Stanford with (from left) Carrie Hunter, Steven Smith, and Emilie Bergman after their release in the fall of 1975. Photograph courtesy of David Hamburg.

Joseph A. Califano, Jr., LL.B., President Jimmy Carter's Secretary of Health, Education, and Welfare, worked closely with the Institute in the Hamburg era, and ultimately was elected an IOM member and a member of the IOM Council. Photograph courtesy of the History of Medicine Division, National Library of Medicine.

Robert Sproull, Ph.D., a well-known physicist and president of the University of Rochester, chaired a committee whose 1984 report advocated the transformation of the Institute into an academy of medicine and led to some particularly painful moments for the young organization. Photograph courtesy of the University of Rochester.

In early May 1977, a *Washington Post* article on the recently released IOM study *Computed Tomographic Scanning* caught President Jimmy Carter's eye. The study urged that hospitals and physicians should not overuse the new technology and that local health planners should approve the installation of new scanners, as well as ensure that they operated efficiently. Carter wanted Secretary of Health, Education, and Welfare Califano to read the article as well.

THE WHITE HOUSE
WASHINGTON

5-3-77

To Califano
Let's take similar
action - stronger if
possible - & include
other devices as
advisable.

J. C.

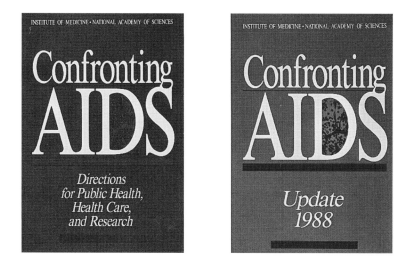

Confronting AIDS (1986) commanded public attention and urged the nation to do more to combat this deadly epidemic. A follow-up report, Confronting AIDS, Update 1988, was used by President George Bush to help formulate government policy toward AIDS.

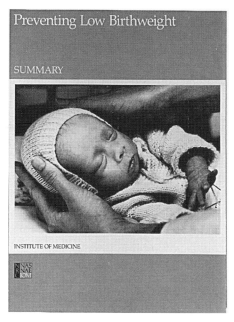

Preventing Low Birthweight attracted a great deal of attention in part because of the dramatic way in which it was released in the spring of 1985 during hearings conducted by Representative Henry Waxman (D-Calif.).

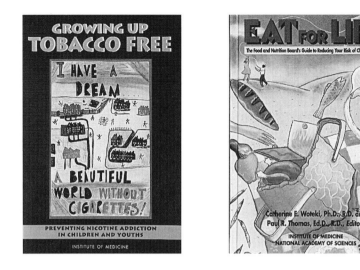

In recent years, the Institute has produced reports that grab a reader's attention visually as well as engaging them with substantive issues, as the covers of these reports show—*Growing Up Tobacco Free: Preventing Nicotine Addiction in Children and Youths* (1994, top left), *Eat for Life: The Food and Nutrition Board's Guide to Reducing Your Risk of Chronic Disease* (1992, top right), and *In Her Lifetime: Female Morbidity and Mortality in Sub-Saharan Africa* (1996, below). Such covers are strikingly different from the more ultilitarian documents that the Institute produced during its early years.

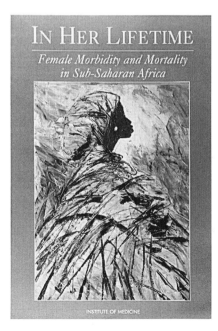

Current National Academy of Sciences President Bruce Alberts, Ph.D. (left), joins past NAS President Frederick Seitz, Ph.D. (right), at the April 1995 dedication of a portrait of Frank Press, Ph.D. Press, a noted physical scientist from the Massachusetts Institute of Technology, had been President Carter's science advisor prior to taking over as NAS president in 1981. Alberts, a well-known biochemist and molecular biologist from the University of California at San Francisco, succeeded Frank Press in 1993. Photograph courtesy of the National Academy of Sciences.

THE WHITE HOUSE

WASHINGTON

December 15, 1995

Greetings and congratulations to all those gathered to celebrate the Institute of Medicine's twenty-five years of devoted public service.

Since 1970, the Institute of Medicine has contributed thoughtful and wise health policy analysis covering an extraordinary range of issues -- from mental health to Medicare, from nutrition to new vaccines -- that are of great interest and importance to us all.

As we continue to work toward a high-quality, fully accessible health care system for all our people, your work is essential to maintaining the finest possible care. I am pleased to commend the Institute for its efforts, day in and day out, to improve the health of the American people.

Best wishes for every success in the years to come.

Bill Clinton

A letter from President Bill Clinton commemorated the Institute's 25th anniversary in December 1995.

5

Expansion of the Institute of Medicine

At the end of March 1987, the *Wall Street Journal* ran a story about the Institute of Medicine (IOM) on its Washington politics page. In contrast to the journalistic pieces on the IOM that had appeared during the confrontation over the Sproull report, this one had a decidedly upbeat tone. "After more than 15 years of comparative obscurity," wrote *Journal* reporter Alan Otten, "the Institute of Medicine . . . seems to be gaining visibility and clout in Washington." The article cited the many problems that had previously characterized the IOM, such as the quick turnover in presidents, the constant scramble for money, and the ambivalent attitude of the foundations, and suggested that these problems had been remedied in the short period between the end of 1985 and the spring of 1987.[1]

No doubt, the newspaper overstated both the degree of the initial problems and the permanence of the solution. It was nonetheless the case that the IOM rebounded from the crisis of the Sproull report under the leadership of Samuel Thier. In a short time, Thier restored the confidence of foundations in the Institute of Medicine, leading to a more stable pattern of financing and ending the yearly struggle to make ends meet. Resolution of the financial issues strengthened the position of the Institute of Medicine within the National Academy of Sciences and paved the way for a major expansion of IOM responsibilities and staff. As a result, the years between 1985 and 1991 were ones of major growth for the Institute of Medicine.

Samuel Thier and the Outreach to Foundations

Thier, a comparatively young man of 48 when he became IOM president, had the sort of leadership style that inspired confidence. Refusing to become bogged down in details or to succumb to cynicism, he simply assumed that he would be able to meet his objectives,

projecting an air of competence that both reassured and energized
people. Among many other positive attributes, Samuel Thier was a
quick study. Even before he took over on a full-time basis, he realized
the need to build a strong endowment that would allow the IOM to set
its own agenda. Turning a disadvantage into an advantage, he used
the Sproull Report as a form of leverage. Because the foundations had
subjected the Institute of Medicine to such a probing and painful
analysis and because the Institute of Medicine had begun to put its
house in order, Thier argued that the foundations now had a moral
obligation to come to the IOM's aid. He hastened to reassure these
foundations that he had absorbed the many criticisms of the IOM and
would take steps to remedy the problems. Under his leadership, he
promised, the IOM would strive to complete its studies more quickly,
with the goal of cutting in half the time necessary to complete studies.
Using the IOM's convening power, he planned to bring together the
parties interested in drug development and health technology
assessment in order to facilitate communication among participants
in these particular fields. Finally, Thier vowed to make the IOM more
visible to the government, so that government agencies turned
reflexively to it for advice.[2]

Like his predecessors, Thier was a respected academic doctor who
had produced more than 80 research papers in his chosen field of
kidney function. More importantly, however, he had early shown a
penchant for medical administration and proved himself to be a
superior clinician. The son of a physician, Thier decided to follow his
father's profession. "My father was a general practitioner and loved
what he was doing and I just thought it was an exciting life," he
explained. Thier graduated from the State University of New York
Upstate Medical Center at Syracuse in 1960 and then entered the
elite echelons of the medical profession by taking his internship and
residency at Massachusetts General Hospital in Boston. By 1966, he
had earned the coveted position of chief resident at Massachusetts
General, which prepared him for a career in academic medicine that
included positions as chief of the renal service at Massachusetts
General and vice chairmanship of the Department of Medicine at the
University of Pennsylvania. In 1975, only 15 years removed from
medical school, he became chairman of the Department of Medicine at
Yale. Fellow doctors admired both his ability to respond to emerging
trends in medicine and his clinical skills in diagnosis and treatment.[3]

By the time Thier arrived at the Institute of Medicine at the end of
1985, he had already played an active role in the organization's affairs
and in medical affairs more generally. Elected to the IOM in 1978,
Thier had chaired the Board on Health Sciences Policy and been a

member of two study committees. Just before coming to the IOM as president, Thier served as chairman of the American Board of Internal Medicine, the national body that set certification standards for doctors in internal medicine and its subspecialties. Under his and William Kelley's leadership, the board had established the subspecialty of geriatrics. In addition, Thier held many appointments related to his academic position, such as chief of medical service at the Yale–New Haven Hospital and member of the editorial board of the *New England Journal of Medicine*.[4]

When Thier presided over his first Council meeting on November 18, 1985, it immediately became apparent that the IOM would no longer be, in the words of a veteran IOM member, "Sleepy Hollow." Thier cut through much of the talk that tended to slow the organization down. Too much energy, he said, had been expended on determining whether the IOM should be involved in health policy or the health sciences. He believed that it was not an either–or proposition, but that instead, the Institute should be responsible for what he described as "the entire spectrum of activities within the National Academy of Sciences (NAS) complex that deal with human health." To make sure this was the case, Thier bargained with NAS President Frank Press to initiate a review of National Research Council (NRC) activities that should be transferred to the IOM. At the same time, Thier realized that if these units were to be transferred to the IOM, the IOM would have to improve its ability to provide oversight on basic science issues by increasing the number of scientists on the IOM Council. Thier also took immediate steps to address the IOM's financial problems. He got Press to agree to raise the amount of NAS support for the IOM from $596,000 to $800,000. With this additional money, no staff cutbacks would have to be instituted. More importantly, he convinced Press to grant the IOM a two-year grace period during which it would not be penalized for reasonable deficits.[5]

As soon as Thier took over on a full-time basis at the beginning of 1986, he made the rounds of foundations and appealed for core support. For many of the foundations, such as the Robert Wood Johnson (RWJ) Foundation, money for an endowment violated their action-oriented view of philanthropy. They wanted to solve specific, pressing problems at a time when people were increasingly turning to private charities, rather than to the government, for such solutions. They had less interest in endowing institutions and giving them a free rein to run their own affairs. With his usual optimism, Thier visited the Commonwealth Fund, the Carnegie Corporation, the Andrew W.

Mellon Foundation, the Dana Foundation, and the Kaiser Family Foundation and tried to convince them otherwise.[6]

Thier realized that the key to the effort was the Robert Wood Johnson Foundation because other foundations that specialized in health and medicine tended to follow RWJ's lead. Of all the foundations interested in medicine, RWJ had the most money at its disposal, enjoyed a close relationship with the Institute of Medicine, and played a key role in the formation of the Sproull committee. Of all the foundations, however, RWJ also had the most stringent rules for the accounting of funds and made very few grants to endow specific institutions. It preferred to run its own fellowship programs, often in conjunction with other institutions such as Johns Hopkins University and the Institute of Medicine, and to take a coordinated approach to the solution of a specific problem such as homelessness. Aware of these constraints, the IOM proceeded to make a request for a $5 million endowment that occasioned considerable controversy among members of the RWJ board and staff. The foundation asked Thier to meet with a small subcommittee of its board in the fall of 1986.[7]

The meeting proved a considerable success. At the end of October, the Robert Wood Johnson Foundation announced that it would make a special award to the Institute of Medicine. The award consisted of a $5 million grant that came with two conditions: The first was that the IOM raise its endowment goal to $20 million, and the second was that the IOM raise $2.00 for every $1.00 it received from Robert Wood Johnson. "The conditions of the grant are fully acceptable to us," Thier replied, noting that there was "no organization to which the Institute is as deeply indebted for its continuing support as it is to the Robert Wood Johnson Foundation."[8]

The IOM used the RWJ grant as the basis for a capital campaign announced in March 1987. By this time, the IOM already had in hand much of the money necessary to match the RWJ grant. By November 1986, for example, the IOM had received $500,000 from the Commonwealth Fund, $1 million from the Andrew W. Mellon Foundation, $1.5 million from the Kellogg Foundation, and even more impressively, $5 million from the MacArthur Foundation. For the first time, the Institute of Medicine would be able to enjoy the benefits of an endowment that would yield substantial income each year.[9]

These significant new sources of income did not exempt the IOM from the usual sort of confrontation with the foundations. The Robert Wood Johnson Foundation, in particular, proved to be a demanding patron. In 1988, for example, it reduced the IOM's budget request for the RWJ fellows in health care policy from a requested $300,000 to $200,000 and refused to contribute toward the indirect costs of running the program. Becoming very angry, Thier accused the foundation of not honoring its previous commitments. As a compromise, foundation officials suggested that RWJ pay half of the indirect costs and that the IOM pay the other half from its endowment funds. As Marion Ein Lewin, the IOM staff member who ran the fellowship program, explained, foundation officials regarded the endowment money as an "unusual, generous grant" that had served as seed money for other grants. The foundation was also firmly against paying indirect costs. Hence, it seemed reasonable to foundation officials that the IOM use some of its RWJ money to help pay for the one IOM program directly associated with the Foundation. Thier refused to see it this way, insisting that the use of the endowment to defray indirect costs had never been a condition of the RWJ grant. IOM Council members expressed a reluctance to "disrupt relations" with a foundation that had been so generous to the IOM for so long. Thier, for his part, did not want to compromise on an important point "in order to placate a powerful donor."[10] In the end, Thier secured an agreement that after a two-year period, the foundation would pay the full costs of the program.[11]

Despite these inevitable disagreements, the fact is that the IOM enjoyed unprecedented success in its outreach to foundations during the presidency of Samuel Thier. One factor in this success was the surge in the stock market and the economic boom that occurred in the mid- to late 1980s. After a sharp recession in the early 1980s, economic conditions brightened. This meant that the value of the securities that foundations held in their portfolios increased, and because the foundations were required by law to spend a certain percentage of their income, the amount of money they awarded also increased. For example, the Kellogg Foundation discovered at the end of fiscal year 1987 that it had "a significant overrun in income" that had to be spent by the end of the calendar year. This made it receptive to a proposal from Frank Press and Sam Thier for an additional $20 million endowment for the NAS and the IOM.[12] The foundation eventually awarded the NAS and the IOM a $20 million challenge grant to support studies in health, education, and agriculture.[13]

Another factor in the IOM's fund-raising success was that the Reagan revolution appeared to have run its course. Because the foundations were no longer quite so panicked about defining their role in an era of shrinking government, they felt able to make grants to their traditional clients such as the IOM. A third factor was the persistence and confidence with which Thier approached the foundations. Unlike Fred Robbins, who tolerated the fund-raising process because he recognized its importance and knew it was expected of him, Thier made fund-raising a priority and enjoyed the experience. He excelled at reassuring the foundations that their money would be well spent and proved adept at creating a bandwagon effect, urging one foundation not to get left behind by another.

Report Dissemination and Expansion of the Institute

One of the items on which Thier spent the IOM's newfound money was the dissemination of IOM reports. From the day he took over as IOM president, he vowed to make an effort to distribute IOM reports in an organized fashion, beginning with the original proposal and extending through the evaluation of the final report's impact.[14] "Dissemination is our current buzz word," said communications director Wallace Waterfall in 1987,[15] referring to such things as a videotape of the symposium on the medical effects of nuclear war that the IOM prepared for distribution to universities and the special efforts that the IOM made to publicize its study on prenatal care.[16]

The prenatal care study, designed to suggest ways of increasing the utilization of prenatal care by mothers early in their pregnancies, began in 1986. Headed by Joyce Lashof, dean of Berkeley's School of Public Health, it followed from the recommendations of the widely acclaimed report from the Fred Robbins's era that had cited increased prenatal care as a primary means of preventing low birthweight. Therefore, the very fact that the IOM undertook the second study indicated a new ability to achieve continuity in its program. To support the study on the utilization of prenatal care, the IOM secured funding from the Carnegie Corporation, the Ford Foundation, the March of Dimes, and the Rockefeller Foundation. To ensure wide dissemination of the final report, which appeared in October 1988, the IOM received additional money from Carnegie and Ford that enabled it to issue a separately bound summary of the report to more than 10,000 people and organizations. On the evening before formal release of the report, the IOM hosted a special dinner for 70 key leaders in the field of maternal and child health, and the following morning the IOM held a press conference. Forty reporters attended, leading to reports on the CBS Evening News and the MacNeil–Lehrer NewsHour and in the *New York Times,* the *Wall Street Journal, USA Today,* and many local newspapers.

The IOM next focused on giving speeches about the report to influential groups and on getting coverage in medical and scholarly journals. These efforts resulted in presentations in such settings as the U.S. Conference of Mayors, the Council on Foundations, and the

National Governors' Association and articles in such places as the *Journal of the American Medical Association* and the *Journal of Social Issues*. In the next phase of dissemination activity, the IOM sought to interest Congress in the report. In the summer of 1988, Sarah Brown, the staff study director, testified before both the House Committee on Government Operations and the Senate Finance Committee.[17]

This sort of attention to dissemination became a key part of Thier's plan to improve the performance and influence of the IOM. As one document of the period stated, "The recommendations of a study can only be regarded as disseminated when they are acted upon by the health enterprise." Other components of the IOM's "new directions" included better targeting of health policy problems as they developed, greater versatility in responding to problems, and closer involvement of members in the creation of the program plan and the execution of IOM studies.[18]

Increased funds and more attention to dissemination brought positive results in the form of increased prominence within the National Academy of Sciences, growth of the IOM staff, and expansion of the IOM's organizational responsibilities. One sign of its increased clout within the larger organization was an agreement that increased the NAS funds that went to the IOM. These funds took the form of payments for the overhead costs of some of IOM's program activities. The more that the IOM contributed to the total pool of overhead funds, the more it could demand from the Academy.[19] In fiscal year 1988, the IOM expended more than $6.5 million dollars in program activities, compared with $3.02 million in 1986. At the beginning of 1988, the IOM had 85 project staff members (compared with 20 in December 1986) and 20 core staff members. The staff was growing so rapidly that finding a place to house them in relative proximity proved a serious challenge.[20]

A key element of the IOM expansion involved the July 1, 1988, transfer of the Medical Follow-Up Agency and the Food and Nutrition Board from the National Research Council to the Institute of Medicine. These organizational changes marked the realization of plans that went at least as far back as the Ebert report in 1982, which had called for the IOM to assume more administrative responsibility for the components of the Commission on Life Sciences (CLS) that dealt primarily with medicine. When Thier took over, he asked Frank Press once again to investigate the organization of the NRC with an eye toward identifying those parts of the Commission on Life Sciences that more properly belonged in the IOM. Press assigned Walter Rosenblith, emeritus professor at MIT and former member of the

Board on Medicine, to head the group charged with this task. His primary recommendation was that the Medical Follow-Up Agency should be transferred to the IOM, and his secondary recommendation called for the Food and Nutrition Board to have part of its activities overseen by the National Research Council and part by the Institute of Medicine. The IOM Council preferred a simpler arrangement, in which the Institute assumed administrative responsibility for the Food and Nutrition Board, with an understanding that the IOM would "assign to CLS . . . those activities that were recommended by [Rosenblith's] committee for retention under CLS oversight."[21] In this case, unlike so many others in the past, the IOM got what it wanted. Both the Medical Follow-Up Agency and the Food and Nutrition Board came to the IOM.

Created in 1946, the Medical Follow-Up Agency brought an epidemiological focus to the Institute of Medicine. The agency's main purpose was to facilitate the use of federal records, primarily those of the armed forces and the Veterans Administration, for medical research. The agency also maintained a special data base on twins who had served in the armed forces. Among topics of interest to the agency, which contained a staff of six professionals, were the psychological and medical results of military captivity, the risk of cancer following exposure to a nuclear weapons test, and the natural history of various forms of hepatitis. An eight-person Committee on Epidemiology and Veterans Follow-Up Studies, headed by Richard Remington, vice president for Academic Affairs at the University of Iowa, supervised the agency's studies which were financed, at the rate of about $1 million per year, almost entirely from federal funds. Unlike the other parts of the Institute of Medicine, the Medical Follow-Up Agency engaged in the conduct of original research, often with statisticians and epidemiologists on the staff acting in collaboration with clinical investigators and epidemiologists from academic medical centers.[22]

The Food and Nutrition Board was a larger and more complex undertaking. Started during the Second World War, it addressed issues of critical importance that pertained to the adequacy and safety of the nation's food supply as well as matters that related to proper diet and nutrition. Among the questions it sought to address were ones of considerable political and economic sensitivity, such as the effects of chemical additives on the quality of the food supply or the nutritional qualities and harmful side effects of particular foods. Most of the members of the 12-person board were academic physicians or scientists concerned with the study of nutrition. Typical of the work of the Board was a large 1989 report on *Diet and Health* that recommended the appropriate daily levels of fats and salt necessary to maintain health and prevent disease. Other aspects of the board's work concerned the international dimensions of nutrition, such as how to correct the deficiency of vitamin A in the diets of Third World children that led to blindness. Still other parts of the board's mission involved the maintenance of proper nutrition during pregnancy and lactation. In 1988, the board's disparate activities cost about $1.5 million to run and required the services of some 14 professional staff members.[23]

Not only did the IOM acquire new agencies as part of its newfound prosperity, it also revived components that had been moribund and invented new entities. The Board on International Health served as a good case in point. An important priority of David Hamburg's, international health had languished as an IOM activity during the 1980s, becoming the IOM division that encountered the most difficulty in attracting outside funds. To the proprietors of an increasingly troubled American health care system, international problems appeared remote, and the Reagan administration was reluctant to spend money on controversial forms of foreign aid such as advice on population control. As a result of these forces, the Board on International Health, which had struggled to come into existence in the first place, was effectively dissolved by the end of 1987. Thier and his staff set out to raise funds for a revived board and succeeded in gaining money from the Rockefeller Foundation, the Public Health Service, and the Agency for International Development. The IOM also contributed some of its newly gained endowment money. As a result, the new Board on International Health met for the first time in January 1989.[24]

In reconstituting the board, Thier vowed to avoid the mistakes of the past. The new board, unlike the old one, would have a firm funding base and be less susceptible to changing political fashions in which health shifted as a priority in economic development and foreign aid portfolios. It would be smaller and better focused than the previous board, and it would include foreign experts on health care policy. With this new outlook, Thier hoped that the new board would shape "a clear image of a distinctive role for itself" and convey this to the international health community.[25] By 1990, the IOM's program in international health included a wide array of projects, such as a study of malaria prevention and control and another of female morbidity and mortality in Africa.[26]

This latter project ultimately resulted in the publication of a 1996 volume that showed how the IOM had changed over the years. Like other complicated projects, this one took a long time to complete. It required almost six years to gather sufficient funding, find an appropriate focus, and perform the analysis. The final report covered sub-Saharan Africa and not the entire African continent. It was nonetheless an ambitious attempt to define the gender-related burden of health problems for females across the life cycle. As with all recent IOM reports, it began with an accessible summary that, for example, presented the main results in the form of two large tables. Furthermore, the book had an attention-getting main title, *In Her Lifetime,* and an arresting cover. Against a striking gold background, an impressionistic picture of an African woman, done in shades of red, orange, and black, appeared. It made quite a contrast to the utilitarian covers, in an institutional shade of yellowish gray, that had graced IOM reports through all of the 1970s and much of the 1980s. The old reports had consisted of photo-offset typescripts with jagged right margins; the new reports contained sleek typefaces and crisp graphics that gave these publications a professional appearance and made them much easier to read.[27]

The chief drawback of studies such as the one on female morbidity in Africa was the length of time required to complete them. As part of the effort to respond more quickly to current concerns, the IOM started a number of informal forums. One was the Forum on Drug Development and Regulation, "a meeting ground for the exchange of ideas and information," that began in July 1986. The idea behind this effort, an experiment that soon evolved into a regular activity of the Institute's Board on Health Sciences Policy, was to provide "regular meetings in a nonadversarial environment for representatives from government, industry, and academia to discuss pharmaceuticals." Recognizing the value of such an institution, federal agencies concerned with drug development such as the Food and Drug Administration, professional organizations such as the American Medical Association, and private trade associations such as the Pharmaceutical Manufacturers Association all contributed to the forum's upkeep. The Forum on Drug Development, as the entity ultimately became known, involved little of the internal clearances and other hindrances that often delayed IOM initiatives. A unique use of the IOM's convening power, it made the IOM privy to the latest developments in the field and led to workshops on related topics, such as a 1990 workshop on the development of drugs for pediatric populations.[28]

A more ambitious effort to respond quickly to current health concerns began in the Fred Robbins's era with a 1983 report that recommended establishing a consortium for assessing medical technology. This idea attracted interest in Congress and led to the passage of the Health Promotion and Disease Prevention Amendments of 1984, which included a provision for the Council on Health Care Technology. A subsequent law, passed a year later, made it possible for this council to be formed as an IOM activity. By the time the first meeting took place in April 1986, Samuel Thier, like Fred Robbins before him, had become an enthusiastic proponent of the concept. The council required a great deal of Thier's attention, because Congress appropriated money for it through the National Center for Health Services Research (NCHSR) and Health Care Technology Assessment (part of the Public Health Service) and made all money contingent on the receipt of private matching funds, at rates that changed over the course of the project. Thier organized a fund-raising campaign in which, for example, he solicited funds from the Medical Products Group at Hewlett-Packard and the Du Pont Medical Products Department, as well as numerous insurance companies. The need to raise funds in such a piecemeal fashion hampered the operations of the council, which was phased out in 1990 in favor of a Committee on Clinical Evaluation. "We believe that the requirements to raise private sector support and to spend it before it can be matched by NCHSR . . . hindered the work of the Council," Enriqueta Bond of the IOM staff explained.[29]

In its relatively short life, the 16-person council, chaired by William Hubbard, the former head of the Upjohn Company, and cochaired by Jeremiah Barondess, a professor of clinical medicine at Cornell University Medical College and later president of the New York Academy of Medicine, studied aspects of the use of technology in health care. Perhaps the most important of its projects, the 700-page *Medical Technology Assessment Directory,* illustrated its role as a clearinghouse of information in the field. Appearing in 1988, this directory provided a list of resources on which hospitals, insurers, and others in the health care field could rely to make difficult decisions related to medical technology. If, for example, a hospital wished to purchase a lithotriptor to destroy kidney stones, it could find listings that would direct it to the available information on this subject. Insurance executives, concerned about the proper use of magnetic resonance imaging, could use the directory to get a quick sense of studies designed to test the cost-effectiveness of such techniques. Clifford Goodman, the council program officer, described the directory as a volume "designed to get dog-eared" and as a step toward creating an effective network of practitioners in the field of health care technology assessment. The completion of such studies, as well as ones on computerized patient records, earned plaudits from Congress. In 1987, renewing the program for three additional years, the Senate

Committee on Labor and Public Welfare, praised the council for providing an "important forum for joint government and private cooperation to promote the development of methods of health care technology assessment."[30]

Although this important council proved short-lived, it showed a new willingness of the IOM to experiment with entities other than study committees in order to influence health care policy—another sign that the IOM had come out of its early 1980s malaise. Indeed, by the end of the decade, many of the problems cited in the 1984 Sproull report had been resolved. The key factor was that the absence of a financial burden gave the IOM a new freedom to set its own agenda. Samuel Thier feared that the National Research Council, with its ethos of service, had become a "job shop" that responded to all requests for assistance. He preferred that the IOM take charge of its own activities, even at the expense of turning down some requests. In this spirit, the IOM began to scrutinize requests far more carefully than it had previously.[31]

At the same time, the IOM contemplated formulating its own requests. The Board on Health Care Services wanted to focus the attention of the Institute of Medicine on the broad question of access to health care services. This topic, with its breadth and complexity, resembled the sort that the Board on Medicine had once hoped to address. In the intervening years, the IOM had discovered, however, that few foundations or government agencies wished to finance studies that appeared to be diffuse and unrelated to political realities. Even though the IOM had managed to secure NRC approval for such studies, it had seldom been able to gather enough money to get them under way. In October 1988, the Board on Health Care Services tried again, proposing that the IOM develop a statement of objectives on this topic, issue an annual status report on health care access in the United States, sponsor a structured debate on critical policy issues, and perform in-depth studies that would generate policy recommendations. Thier, for his part, recognized that the topic was of interest to a broad segment of the IOM membership.[32]

Although the IOM never carried out the ambitious program suggested by the Board on Health Care Services, the theme of access became central to the IOM's operations in 1989. "The quality, access, organization, and financing of preventive and clinical services," emerged as the first of six cross-cutting themes that the IOM used to describe its program and identify its priorities. Identification of the themes by the Program Committee marked an explicit effort to "bind program efforts together and bring cohesion to disparate efforts." The annual report for 1989 included a chart that matched each of the

IOM's many projects with one or more of the cross-cutting themes. The Division of Health Promotion and Disease Prevention's project on a research agenda on aging, for example, matched theme number three, "research opportunities and resources." The Food and Nutrition Board's study of nutritional status during pregnancy and lactation fell under theme five, "critical health issues, including those of special populations, defined by gender, race, age and socioeconomic status."[33]

The formulation of cross-cutting themes indicated that by the end of the 1980s, the chief problems of the IOM had become controlling growth, not soliciting funds. At the end of 1987, Thier turned his attention away from fund-raising and toward strengthening the IOM program. Twenty million dollars was all the organization could reasonably be expected to raise, he told the IOM Council. With fund-raising no longer the top priority, Thier set four goals for himself. First, he wanted to develop a clear set of program objectives, as in the six cross-cutting themes. Second, he hoped to create a work plan that had a span of at least two years to enable the organization to address larger, more complicated problems. Third, he expected to continue his interactions with key officials in the executive branch and in Congress so as to ensure the continuing influence of the IOM on health care policy. Fourth, he sought to achieve one of the long-standing objectives of the IOM, which was to increase members' participation in the IOM's affairs.[34]

Even as Thier's focus began to shift, each year, it seemed, brought more staff and a larger endowment. When the Food and Nutrition Board and the Medical Follow-Up Agency became parts of the IOM, for example, they raised the number of staff members to 132. By July 1988, the Institute of Medicine had become the largest single budget item in the National Academy of Sciences complex. At the beginning of 1989, the IOM had more than 150 employees, project expenditures of $12.7 million, 60 projects under way, and 20 reports scheduled to be released that year. By the end of the year, it had raised more than $23.2 million from its capital campaign, including $18.9 million in endowment.[35] Whereas before the IOM Council had spent an inordinate amount of time trying to come up with a financial development strategy, it now thought of ways to control the consequences of growth. At one point, for example, Thier suggested the development of a computer program that would enable one division to track the activities of another through the use of key words.[36]

Samuel Thier credited his staff members with making much of the growth possible. When Charles Miller retired as executive officer at the end of 1988, Enriqueta Bond took his place. As Roger Bulger, Karl

Yordy, and Charles Miller had done before her, she managed the IOM's increasingly complex internal affairs, operating with an efficiency that enabled Thier to reserve his energies for the four goals he had set for himself. She received capable support from a team of administrators that included individuals who had worked at the IOM since the days of John Hogness, such as administrative officer Louis Cranford, membership services director Jana Surdi, and communications director Wallace Waterfall.[37]

The Institute's Program

Financial security and administrative efficiency defined means, rather than ends. Although they ensured the organization's survival, more money and tight management implied nothing about the organization's future direction. The IOM's identity continued to be bound up with its program. Even with the influx of money and the concentration on report dissemination, only a few studies issued by the IOM in the late 1980s or early 1990s, such as those on prenatal care and on nutritional guidelines, captured the attention or imagination of the general public. Instead, the organization continued to speak out on a wide variety of medical issues and to influence, in a limited way, the direction of public policy. Each project and each study committee had to balance the demands of responsible science and the exigencies of public relations in order to respect the limits of the evidence and still communicate a message.

The IOM functioned best when it used the technical expertise of its members to address a well-defined area of public policy. The best example concerned the development of vaccines, a topic that was a continuing concern of the organization at least since the era of David Hamburg and one to which Fred Robbins had devoted much time and attention. In 1985, as a result of Robbins's efforts, the IOM issued two important reports on this subject, *Vaccine Supply and Innovation* and *New Vaccine Development: Establishing Priorities*. Three years later, *An Evaluation of Poliomyelitis Vaccine Policy Options* appeared, which updated the work on this subject that the IOM had done in 1977. The 1988 polio report recommended the use of an injectable vaccine that could be given in combination with other childhood immunizations, a recommendation that was adopted by the Public Health Service's advisory committee on immunization practices.

The report on *Vaccine Supply and Immunization,* which appeared at the end of July 1985, had a definite effect on public policy. It contained a warning that the supply of vaccine was "precarious" and posed a "threat to the public's health." The problem was that manufacturers hesitated to make the vaccines for fear of being sued by someone who had been harmed by them and having to pay exorbitant damage claims. The solution, according to the group headed by Jay Sanford, dean of the Uniformed Services University of the Health Sciences, was the adoption of a new compensation system that would relieve the manufacturers of the uncertainty they faced and assure adequate compensation to victims of vaccine-induced injury. The commission also saw the need for a group that would continue to monitor the situation and recommended the creation of a national vaccine commission for this purpose.[38]

The IOM made sure that the report was distributed widely, particularly to the people who made public policy that affected vaccine development. It also commissioned a science writer to produce a nontechnical summary of the report that better enabled the media to understand and report on the study, and it briefed congressional staff members on the report. These efforts paid off in March 1986 when Representative John Dingell (D-Mich.), chairman of the House Committee on Energy and Commerce, and Senator Orrin Hatch (R-Utah), the chairman of the Senate Committee on Labor and Human Resources, requested that the IOM convene a workshop to refine the report's conclusions for possible use in the legislative process. Financing the workshop through special dissemination funds from the Carnegie Corporation, the IOM complied with the request. The workshop led to a bill that proposed a national vaccine program and an injury compensation system, which Congressman Henry Waxman (D-Calif.) introduced in July 1986. President Reagan signed a modified version into law on November 14, 1986, as part of a larger health measure. In this manner, the Institute of Medicine came to be the author, if such a thing were possible in the collaborative world of Congress, of the National Vaccine Program and the National Childhood Injury Act of 1986 that were contained in the law.[39]

The report on vaccine innovation and supply illustrated nearly all of the elements of Samuel Thier's strategy to increase the IOM's effectiveness. First, the IOM paid a great deal of attention to the process of dissemination, making sure the report came to the attention of key policymakers and journalists. Second, the IOM responded quickly to Congressman Dingell's and Senator Hatch's request for assistance. In addition, the IOM displayed a sense of flexibility throughout the process. In the past, the organization might simply have issued a report and left it for others to interpret. In the Thier era, the organization did not hesitate to get involved in the finer details of politics and, in effect, to collaborate with congressional staff in writing the bill. Where John Hogness might have shied away from congressional testimony and direct involvement in the course of legislation from bill to law, Samuel Thier appeared to delight in these things. For him, passage of a law that bore the IOM's imprint represented an organizational triumph.

A less dramatic triumph came from the work done by the IOM on Reye's syndrome. Acting in an oversight capacity, the IOM issued a series of short reports, all highly technical in nature, on the Public Health Service's conduct of a study of the association between analgesics, such as aspirin, and Reye's syndrome. These reports appeared between 1984 and 1987, simultaneously with the formulation and the conduct of the study. Through these reports, the IOM called attention to an issue that required Food and Drug Administration (FDA) intervention. As a result, the FDA ordered that aspirin packages carry special warnings advising parents not to give the product to children with influenza or chickenpox. This public health measure might not have been undertaken without IOM's intervention. It was the sort of thing that the IOM listed in its annual reports under headings such as "making a difference" or "impact."[40]

Not everything worked so well, as IOM's experience with the Committee on Health Objectives for the Year 2000 demonstrated. This project was larger, more diffuse, less well-suited to the National Academy of Science's ethos of scientific investigation, and much closer to the surface of partisan politics than others that the IOM undertook. It involved a process that had begun in the Carter era and led to the 1979 publication of *Healthy People*. Issued by Surgeon General Julius Richmond, the book had established broad goals for improvement of the health of Americans. Around 1986, the Public Health Service began the long process of producing new goals for the year 2000 and requested the IOM's help in the procedure. The IOM's experience in the Carter years with this sort of exercise had been painful, leading to major disagreements over the relative emphasis on individual responsibility and government obligation in maintaining the nation's health. Nonetheless, the IOM consented to join the new effort. Unlike the Carter initiative, this one was supposed to come from the bottom up rather than the top down. Rather than convening a group of national experts in Washington, D.C., the Public Health Service hoped to hold hearings in locations across the nation, in order, in Enriqueta Bond's words, to "create a sense of ownership" by the local agencies and private-sector organizations that would have to implement the new objectives. The role of the Institute of Medicine was to help organize the process by holding regional hearings, summarizing the results in a form that would be useful to the Public

Health Service, forming a consortium of professional and voluntary organizations and soliciting its input to the process, and convening a conference to launch the new objectives.[41]

The Institute of Medicine appointed a steering committee for the project, which conducted regional hearings at the beginning of 1988. The idea was to hold the hearings before the presidential election distracted public attention from the process. The hearings produced a series of grassroots laments that were difficult to synthesize into a larger and more coherent list of objectives. In Birmingham, Alabama, an academic physician from the University of Tennessee testified on the subject of breast cancer, a disease that he characterized as "common, curable, and easily detectable." Despite this fact, people continued to die of breast cancer, most often because of a delay in diagnosis that was the result of "negative feelings about health care facilities, avoidance and denial and societal conditions that lead to a fatalism about the disease." A physician from Nashville, Tennessee, said that poor people delayed seeking treatment for high blood pressure because of financial barriers and also because of a "culturally determined belief that favors reliance on family, friends, and spiritualists over medical professionals." A member of the Washington State House of Representatives told the IOM study group in Seattle that the number one issue was health care financing. "Legislators," he testified, "see health dollars as a kind of pac man (a popular video game of the period) eating various bits of money . . . we're seeing costs going up and access going down." Just how these problems could be addressed through the goal-setting process was not immediately evident.[42]

As the hearings took place, IOM staff sat in on discussions at the Public Health Service in which federal officials critiqued drafts of the health objectives for the year 2000. From the IOM perspective, the drafts often showed what one staff member described as "political accommodation rather than scientific rationale." The staff member questioned the need to add a category on the "safety of foods, drugs, and medical devices," which appeared to be a means of getting the FDA involved in the objectives process and nothing else. He claimed that the addition of category 22 for people with disabilities "seems to be the result of strong political pressure to do something for the disabled."[43] In other words, the Institute of Medicine believed that setting health goals should be a scientific process, rather than one dominated by bureaucratic politics and the needs of minority groups to advance their political agenda. IOM staff members felt uncomfortable with the pluralism of the policy process, preferring instead to seek consensus among well-trained and experienced health professionals. Holding such a view, the Institute of Medicine, or more properly its staff and steering committee concerned with this project, found it difficult to participate in the making of policy from the bottom up. Because such a policy model was antithetical to the National Academy of Science's very nature, any IOM reports that reflected this view would, in all likelihood, not be approved by the NRC.

The conference that was to be the culmination of the project took place in the spring of 1990. By this time, George Bush was President of the United States. Because the Public Health Service hoped the President would deliver the keynote address at the conference, planning for the conference became complicated and politically contentious. The Public Health Service, for example, invited the United Way to cosponsor the conference, providing a link between the health objectives and the private sector and enabling the United Way to raise funds for the conference. Saying that this arrangement violated National Academy of Sciences rules, Samuel Thier wanted the IOM to drop out as a cosponsor. To navigate around these difficulties, the Public Health Service and the Institute of Medicine negotiated an agreement under which the Year 2000 Health Objectives Consortium, created and managed by the IOM, became a cosponsor with the United Way. The wording of the objectives generated even more discussion. The White House wanted to keep the President away from controversial matters; hence, the objectives related to family planning and teenage parenthood were vague on the use of contraception by teenagers, the provision of sex education, and the subject of abortion. In the end, the President did not appear at the September 1990 conference, which featured the release of *Healthy People 2000: Citizens Chart the Course,* the IOM publication that summarized the testimony it received. Issuing a proclamation, President George Bush sent Secretary of Health and Human Services Louis W. Sullivan to speak in his absence. Although the surgeon general described the report as the "cornerstone" of the process, the collaboration between the IOM and the Reagan and Bush administrations involved considerable strain and misgivings.[44] In this case, it was not clear how much of a difference the IOM had made.

The IOM fared better when it had more control over the process, as in a 1988 study of the future of public health. This study resembled the one on setting goals for the year 2000 in that it consisted of predicting the future course of health policy and making suggestions to alter this course for the better. Both studies involved the process of seeking input from local health practitioners and professional organizations in order to arrive at some sort of consensus. The difference was that in the public health study, the IOM did not have to reflect grassroots opinion but could rely instead on the advice of its steering committee, which contained such luminaries and IOM members as Richard Remington, vice president for academic affairs at the University of Iowa; David Axelrod, commissioner of health for the State of New York; Lester Breslow, a professor of public health at the University of California at Los Angeles; Melvin Grumbach, head of the Department of Pediatrics at the University of California at San Francisco; and Robert J. Haggerty, president of the William T. Grant Foundation. These individuals shared a set of assumptions, for example, that the maintenance of public health constituted an important mission of government at all levels and that public health officials should have the advantage of expert knowledge derived from fields such as epidemiology and biostatistics rather than being forced to make decisions on the "basis of competition, bargaining and influence." In other words, these people, experts themselves, believed that professional expertise should inform politics.

The report sought a reaffirmation of the nation's traditional public health mission. The committee found that the nation had "slackened [its] public health vigilance and the health of the nation is unnecessarily threatened as a result." Signs of disarray were everywhere. There was little agreement on the public health mission, with the result that public health activities had become fragmented and politicized. Mental and environmental health, for example, were often separated from the rest of a state's public health activities. Morale among public health officers was low and turnover high, with the increasing substitution of political appointees for public health experts. The committee believed that public health departments at the state and local levels should return to the activities they had let disintegrate, such as gathering statistics that described the health of the community. The committee recommended that every state should have a department of health, headed by a cabinet-level officer and distinct from the department of income maintenance, as well as a health council. Through these organizations, according to the report, states should put in place a series of disease control measures that responded to modern conditions such as AIDS, cancer, and heart disease. The report concluded with the message, "Public health is a vital function that is in trouble. Immediate public concern and support are called for in order to fulfill society's interest in assuring conditions in which people can be healthy."[45]

The ringing rhetoric demonstrated the desire of the Institute of Medicine to make itself heard. Most readers probably did not read the entire 217-page report, preferring instead to consult the 18-page summary, which was attractively bound in a softcover booklet with a salmon and turquoise cover. In the past, IOM reports, written in bland and cautious prose with little regard for the felicities of language, had been dauntingly difficult to read. Only the most devoted could make their way through them. The essence of the recommendations mattered, not the language in which they were expressed. By way of contrast, *The Future of Public Health* read as though it had been written by a professional writer, although it was in fact the work of the IOM staff. At times, the recommendations were vague, as in the exhortation that "agencies must take a strategic approach, developed on the basis of a positive appreciation of the democratic political process." However, the report could always be understood.[46]

In a certain sense, clarity, aided by an orderly dissemination process, translated into influence. This report served as a basis for bills that were introduced in state legislatures. The bills sought to revamp the structure and responsibilities of state public health agencies to bring them in line with the IOM recommendations. The report also provided the impetus for a plan developed by the Public Health Service to strengthen public health in the United States. In other words, the project became the focal point for a generation of reform efforts in the field of public health.[47]

Serving as a catalyst for action was a congenial role for the IOM, not only on large global studies such as the future of public health, but on smaller, more specialized studies as well. Typical of these was a report entitled *The Computer-Based Patient Record* that appeared in 1991. The message of this report, the work of a 10-person study committee chaired by Don Detmer, a professor of surgery and vice president for health services at the University of Virginia, was contained in a seven-page summary. In keeping with the IOM's modern sense of style, the report summary came bound in an attractive booklet with a colorful cover, whose sheer cheerfulness tended to counteract the utilitarian nature of the subject under discussion. For harried readers who lacked the time for the summary, the essence of the report could be gleaned by reading the text box on page six that listed the report's seven major recommendations. The chief recommendation was that "health care professionals and organizations should adopt the computer-based patient record . . . as the standard for medical and all other records related to patient care."[48]

Throughout this report, the study committee emphasized its vision of the future, rather than its reading of the present. Hence, the report, despite its prescient quality, had some of the "gee-whiz" attitude common to studies that described future trends in technology. In the committee's view, the patient record of the future was a wondrous thing, complete with "links to administrative, bibliographic, clinical knowledge and research databases"; decision support systems; "video or picture graphics . . . and electronic mail capability." For practitioners of health services research, such a record, with its rich sources of data that could be gathered and analyzed easily, was a particularly tantalizing prospect. Missing from the report, however, was a coherent sense of how to get from here to there because little in the report connected present-day realities to future possibilities. In the real world, computers broke down; proved incompatible from one part of a hospital to another; required constant maintenance, replacement, and updating; and demanded more knowledge of their operations than users often had. Faced with a new technology that took time to learn, many physicians responded by scribbling notes on charts for others to decipher or dictating reports for others to transcribe. The IOM report, in other words, described a future goal suggested by the potential of technology but did not offer much evidence to show that the goal would ever be reached. At issue was the ability of technology to change behavior, an issue that the report did not address. Nonetheless, it did spur further activity in the field, for example, through the founding of the Computer-Based Patient Record Institute to develop appropriate standards.[49]

As the IOM gained influence, it received numerous requests from Congress to examine controversial matters and help in the adjudication of contentious issues. In these instances, the organization did not have the luxury of simply rallying the health care establishment in support of future goals. Instead, it had to confront the limits of the influence of science on public policy.

A good case in point concerned the mandate of the 99th Congress that the IOM investigate the subject of the health care received by homeless people. As the report, issued during the fall of 1988, entered the NRC review process in the winter of 1987–1988, it encountered numerous objections and difficulties. At the heart of these objections was the concern of NAS President Frank Press that the committee had gone beyond the boundaries set by the evidence and strayed into areas not contained in the congressional mandate. Bruce Vladeck, chairman of the study group who later became the head of the Health Care Financing Administration during the Clinton presidency, told the committee that he saw himself as its advocate in the review process.[50] After repeated delays and substantial NAS editing of the report, Vladeck complained that the edited report lacked "emphasis, strength, and passion." To counteract this impression, he drafted a

"Chairman's Message" that contained very strong language. It described homelessness as "a scandal and an outrage" that demanded not only "dispassionate analysis" but also "a more direct and less qualified scream of pain and anger." The report, according to Vladeck, contained "lots of analysis but no poetry. . . . We tried to make our report as dispassionate as the IOM/NAS process requires, but the reality we experienced cries out for passion."[51]

In the end, little of Vladeck's passion seeped into the final report. Careful analysis triumphed over soaring poetry, even in an era in which dissemination of results and impact on the policy process were among the highest IOM values. The report appeared with no supplementary statement from Vladeck, who threatened to write about the matter in an op-ed piece for the *New York Times*. Instead, it contained the unexceptionable conclusions that a high level of mental illness characterized the homeless community and that the homeless required stable residences and sufficient incomes. In this regard, the report could be parodied in a manner similar to a sketch from *Beyond the Fringe:* What the homeless needed was not to be homeless. As for the health care received by the homeless, the report suggested that it was inadequate and formed part of a larger problem of providing access to health care for those unable to pay. In the meantime, public policies toward the homeless were fragmented and often ineffective.[52] Perhaps homelessness, unlike the improvement of medical care records, was simply too large and amorphous a problem for the IOM to solve.

If Congress liked to present the IOM with issues for which no solution was imminent, it also requested the IOM's assistance in matters that were simply too controversial for politicians to handle. Such a case occurred during the early 1980s in the Social Security programs designed to pay benefits to people with disabilities. During the 1970s, Congress had grown concerned over the rapid growth of these programs and had sent signals to the bureaucracy to reduce the size of the rolls. When the Social Security Administration responded by removing many people from the rolls in the early 1980s, Congress shifted direction and expressed concern for people taken off the rolls in the midst of a severe recession. Congress passed a law in 1984 that made it more difficult to remove a person from the disability rolls. Certain issues, however, were too difficult for Congress to decide, such as how much credence should be given to complaints of pain among applicants for disability benefits. In response, Congress mandated the creation of the Commission on the Evaluation of Pain and directed the commission to work in consultation with the IOM. The commission, in turn, recommended further study by the IOM, an assignment that the IOM accepted in 1985 and that led to the report *Pain and Disability* in 1987.[53]

An interesting blend of medical doctors and social scientists served on the committee, which was chaired by Arthur Kleinman, a professor of anthropology and psychiatry at Harvard, and David Mechanic, the Rene Dubos Professor of Behavior Sciences at Rutgers. From the beginning, this group recognized that the phenomenon of pain had both a physiological and a behavioral component. The report stated bluntly that "very little is known about the mechanisms underlying such common clinical problems as low back pain" and that "no direct, objective way to measure pain" existed. Rather than dismiss an applicant's protestations of pain, the group recommended that the Social Security Administration should instead assess what the experts called the "functional capacity for work." At the same time, the committee suggested that such things as "chronic pain syndrome" not be added to the list of impairments that automatically qualified a person for disability benefits. The Social Security Administration took the IOM's advice and instituted an early functional assessment of claimants with a primary pain complaint. To be sure, the disability programs remained controversial, particularly during the tremendous surge in the rolls that occurred after 1984. Nonetheless, the IOM had successfully intervened in, and probably improved, a significant component of disability policy at a time when Congress was reluctant to become involved.[54] Furthermore, it had done so by bringing together leading medical doctors and social scientists in ways that were unique to the IOM.

Even as the IOM undertook special assignments for Congress, it also tended to its traditional concerns. In 1990, for example, one of the cross-cutting themes of the Institute's work was "the role, education and supply of health professionals." In support of this theme, the IOM released a report on financing the graduate medical education (GME) of primary care physicians in ambulatory settings. This project came to the IOM not by way of the government but rather through the efforts of two professional organizations concerned with health, the Association of Program Directors in Internal Medicine and the Ambulatory Pediatrics Association. The IOM obtained support for the project not primarily through the government but rather from the Josiah Macy, Jr., Foundation.

The final report lacked the elegant writing and presentation that marked so many IOM reports during this period, in part because it concerned hard-core matters of interest only to those who understood the intricacies of health financing policy. In support of increasing the number of doctors who worked in primary care fields in outpatient settings, the report recommended that technical adjustments be made in the way the Medicare program financed graduate medical education and the way it reimbursed the services of physicians. The report also urged medical schools to reorder their priorities so that primary care teaching and curriculum development were rewarded. As the report noted, physicians in primary care practice responded to community health needs in settings that ranged from solo practice to HMOs; the hospital was no longer a suitable principal focus for the GME experience of such physicians. The committee concluded that "the care provided by future generations of primary care physicians would be enhanced if the GME experience placed greater emphasis on training in primary care outpatient settings." Although the report's effect on public policy was scant, it helped call attention to the need to match the educational system with the realities of medical practice. It received attention from medical schools that were in the process of revising their primary care residency programs, if not from the Congress.[55]

The Institute of Medicine produced a dizzying array of reports and studies in the Thier years. In the field of health manpower and education, for example, the work of the Institute extended beyond primary care to include multiple studies in the fields of geriatrics, allied health services, and occupational and environmental medicine.[56] Often, one study exposed gaps in knowledge or highlighted opportunities that led to the IOM's being asked by a sponsoring agency to do a follow-up study.

At the same time, the IOM responded to numerous requests for assistance from government agencies on studies or projects that the agencies were undertaking themselves. To cite one example, the Institute of Medicine provided scientific advice and oversight for several Centers for Disease Control (CDC) studies of the health of Vietnam veterans. In particular, the CDC wanted to know if exposure to Agent Orange and other herbicides in South Vietnam led to a greater risk of such cancers as Hodgkin's disease and non-Hodgkin's lymphoma. In 1990, the IOM issued a summary report in which it concluded that the CDC Selected Cancers Study made a "useful and important contribution to understanding the relationship between the Vietnam experience and the cancers under study." Such validation was important because an earlier study that had been canceled led to a charge from Representative Ted Weiss (D-N.Y.) that the Reagan administration had deliberately obstructed the study. In such a contested area of public policy as the effects of Agent Orange, with millions of dollars in damages at stake, the public had to be reassured that scientists who performed the investigations were above political suspicions. The IOM helped to supply this reassurance and improve the conduct of the study.[57]

Despite the fact that the IOM did so many studies on such widely disparate topics, the organization moved with considerable caution. If anything, the IOM moved more cautiously in the Thier years than before, in part because of the realization that its actions had real policy consequences and in part because of a new financial independence that made it less dependent on external funders. When, for example, the IOM convened a group to consider what projects, if any, the organization should pursue in the area of medically assisted technology, Thier warned that it should not become enmeshed in the "sociopolitical arguments" that had arisen over medically assisted conception. Instead, the IOM could maintain its neutrality by focusing on the "science base" of this matter. As a result, the Institute decided to sponsor a conference on reproductive biology. The Board on Medicine and the early IOM, one might surmise, would have engaged the question differently.[58]

In these various ways then, the IOM conducted its activities in the years between 1986 and 1991. At times it responded to congressional mandates and, at other times, generated its own funds for studies that had been suggested by the IOM members. Some studies came at the request of government agencies and others at the request of private foundations or professional groups. In some instances, the IOM tried to condition future practices; in others, it sought to redress past grievances. It handled large and amorphous topics such as the future of public health, as well as smaller and more select topics such as the role of pain in Social Security disability determinations. At times, the IOM acted as an agent of reform. At other times, it functioned as an arbiter of public policy or an entity that validated the actions of others. Always, however, the IOM used its special prestige and status, its membership drawn from a wide variety of medical fields and disciplines, and its ties to the National Academy of Sciences to convene an appropriate and distinguished panel to examine the particular question under discussion. In these regards, the Thier era was not unlike the eras that preceded it; the difference came primarily in the increased scope of the IOM's activities.

Organizational Routines

Despite all that happened between 1986 and 1991, questions that dated back from the founding of the IOM still preoccupied the staff and the Council. The issue that most animated the IOM staff concerned the overlap between the missions of the IOM and the National Research Council. Even with the transfer of new agencies from the NRC to the IOM, the boundary between these two parts of the Academy complex was unstable. In November 1989, Samuel Thier complained, as had his predecessors, that the overlap was "not well controlled" and that such topics as animal research, radiation effects research, and environmental studies and toxicology led to duplication and competition between the Institute of Medicine and the Commission on Life Sciences. Potential for conflict on behavioral topics also existed with the NRC's Commission on Behavioral and Social Sciences and Education.[59]

The question that appeared to be of most interest to the IOM Council involved who should be elected to the IOM in a given year, because each year the longest discussions concerned membership criteria. In January 1988, for example, David Challoner appeared before the Council to discuss the final ballot for the 1988 election and the proposed ballot for the 1989 election. At this point in the organization's history, membership elections had become quite structured, with quotas in seven different categories. The largest of these categories was the one broadly based in the medical disciplines; the smallest included physical scientists, mathematicians, and engineers. In 1988, the Membership Committee arranged the quotas to yield a total of 40 new members. To assist voters, the Membership Committee, with the help of IOM members themselves, ranked the nominees in each of the categories. Those ranked highest usually gained membership. In addition, the IOM hoped to elect 10 people directly to senior membership. For the 1989 election, the IOM Council decided to elect 40 new members, as well as 5 senior members and no more than 10 foreign associates. Apportioning quotas among the categories always created disagreements. In 1988, for example, Alexander Capron, a lawyer from the University of Southern California and a Council member, expressed concern that the number of members in the social sciences category was decreasing. Thier replied bluntly that the IOM was "not a representative group and some disciplines and areas of expertise would not be as well represented as others." The idea was not to spread the membership across fields but rather to identify "outstanding people who can contribute to the Institute's task."[60]

As the Institute added new members each year, it bumped up against the limit of 500 members that was contained in the Charter, which increased the pressure to include individuals from specialties not well represented and made the competition across membership categories even more intense. The trouble stemmed in part from the composition of the various membership sections that suggested new members. These sections were narrower in scope than the membership categories. Section Eight, for example, represented family medicine, primary care specialties, emergency medicine, physical medicine and rehabilitation, and osteopathy. It contributed suggestions for those nominated under Category Three, medical disciplines, in the general election. Although family medicine was well represented in this section, it contained few practitioners from the other medical fields it was designed to include. If a section was dominated by people from one field, it was difficult for someone from another field to get nominated or, if nominated, to receive a high ranking from the section and the Membership Committee. This feature of the election process limited the organization's ability to face new problems that demanded the expertise of people in fields that were not well represented in the membership. In addition to these questions related to discipline, the Institute faced other concerns related to the composition of its membership, such as ensuring the election of women and minorities. In a setting in which the organization received 200 nominations for only 40 places, achieving diversity proved difficult. Members tended to rely on what one described as "the old-boy, old-girl network."[61]

To ease these problems, the Council decided in November 1989 to raise the ceiling on membership from 500 to 600. This would permit the Institute to elect 45 rather than 40 members each year. The rationale for the change was that in the past, it had been necessary to control the quality of the membership by limiting the size of the organization. Because of the rigorous procedures followed by the Membership Committee, however, the Council felt that quality concerns had become integral to the nominating process. Hence, the number of members could be expanded, although not without limit, and the quality of the membership could be maintained.[62]

If there was a major change in membership over the years, it concerned the increased medical, as opposed to public policy, expertise of members. There were fewer generalists, more specialists. When the Board on Medicine started, it included individuals such as Adam Yarmolinsky, who were distinguished in their own fields, lacked direct experience in medicine, and became quite interested in medical affairs in large part because of their involvement with the IOM. In the

IOM's first years, it also had members who were regarded as important to the ongoing discussion of national health insurance, people such as Robert Ball and Wilbur Cohen, two Social Security experts who had played major roles in the creation of Medicare. As national health insurance faded as a topic of concern, people similar to Ball and Cohen, who were neither physicians nor health care professionals working in an academic medical center, had more difficulty being elected to the IOM. With the tendency to more closely merge the medical professions and the IOM came the danger of losing sight of the connection between medical policy and larger social concerns.

Despite this danger, many IOM members insisted on preserving its social, as opposed to scientific, mission. In 1989, Samuel Thier noted with pride that Dr. Louis Sullivan had been nominated by President-elect George Bush to serve as Secretary of Health and Human Services. Sullivan became the first person to serve as Secretary of Health and Human Services while already an IOM member. It gave the organization a sort of "in" with the department most concerned with social policy that it had not enjoyed since the days of Joseph Califano. It increased the chances that the department and the IOM could act in concert in facing such issues as AIDS, homelessness, and crack and cocaine addiction. With each of these social problems, it was the IOM that showed the most interest in them among the components of the National Academy of Sciences. Even as the IOM increased its science component, therefore, it remained the branch of the National Academy of Sciences most dedicated to bringing research to bear on modern social problems.

What most distinguished the years between 1985 and 1991 was not a change in the IOM program as much as the experience of phenomenal growth. No longer could the IOM be regarded as marginal to the NAS, an experiment that could be suspended with cause. Instead, the IOM became the branch of the NAS with the most financial resources at its disposal and, in many respects, the most active membership. The Thier years assured the IOM of its survival and made it a permanent entity in the NAS and on the larger national scene.

It even looked for a time as though Samuel Thier would become the first two-term president of the NAS. When his first term ended in the fall of 1990, he began a second. In May of 1991, however, he announced that he had received an offer to be president of Brandeis University that he felt he could not refuse. When Thier left for Brandeis in the fall of 1991, Stuart Bondurant, dean of the University of North Carolina Medical School at Chapel Hill, became the acting president of the IOM. Once again, the IOM awaited a new leader. Unlike Samuel Thier, for whom expectations were low in the face of threats to the IOM's very existence, the new IOM president would lead a large, complex, and growing institution that was expected to play a major part in the nation's health policy. As the *Wall Street Journal* reported in 1987, the Institute of Medicine had achieved considerable clout on the Washington scene, and Samuel Thier had played no small part in this.

Notes

1. Alan L. Otten, "When Institute of Medicine Speaks, People Listen, Because Newest President Won't Let It Be Ignored," *Wall Street Journal,* March 31, 1987.

2. Gina Kolata, "New Directions for the IOM," *Science,* 230 (November, 1985), pp. 524–525.

3. "Executive Profile: Samuel Thier," *Pharmaceutical Executive,* August 1988, pp. 28–40.

4. "Thier to Become Fifth President of the Institute of Medicine," IOM Press Release, September 30, 1985, Yordy Files, Accession 91-051, Institute of Medicine (IOM) Records, National Academy of Sciences (NAS) Archives.

5. IOM Council Meeting, Minutes, November 18, 1985, IOM Records.

6. IOM Council Meeting, Minutes, March 17, 1986, IOM Records.

7. IOM Council Meeting, Minutes, July 21–22, 1986, IOM Records.

8. Samuel Thier to Leighton Cluff, October 21, 1986, and Cluff to Thier, October 21, 1986, both in Yordy Files, Accession 91-051, IOM Records.

9. "Institute of Medicine, Aiming for Endowment, Gets Johnson Foundation Pledge," IOM Press Release, March 31, 1987, IOM Records; IOM Council Meeting, Minutes, November 17–18, 1986, IOM Records.

10. IOM Council Meeting, Minutes, April 18–19, 1988, IOM Records.

11. IOM Council Meeting, Minutes, July 21–22, 1988, IOM Records.

12. IOM Council Meeting, Minutes, July 23–24, 1987, IOM Records.

13. IOM Council Meeting, Minutes, July 21–22, 1988, IOM Records.

14. IOM Council Meeting, Minutes, November 18, 1985, IOM Records.

15. Alan L. Otten, "When Institute of Medicine Speaks," *op. cit.*

16. IOM Council Meeting, Minutes, April 20–21, 1987, IOM Records.

17. Sarah S. Brown, "Final Report of Dissemination Activities for *Prenatal Care: Reaching Mothers, Reaching Infants,*" Yordy Files, Accession 91-051, IOM Records; Institute of Medicine, *Annual Report 1987, Program Plan 1988* (Washington, D.C.: National Academy Press, 1988).

18. "New Directions for the Institute," June 1986, Yordy Files, Accession 91-045, IOM Records.

19. IOM Council Meeting, Minutes, January 27, 1987, IOM Records.

20. Samuel Thier to Members of the Institute of Medicine, December 9, 1987, Yordy Files, Accession 91-051, IOM Records; IOM Council Meeting, Minutes, April 20–21, 1987, IOM Records.

21. "Minutes of the Discussion of the Report of the Study Committee to Review the Programs of the Institute of Medicine and the Commission on Life Sciences," January 26, 1988, IOM Records.

22. IOM Council Meeting, Minutes, January 9–10, 1989, IOM Records; Institute of Medicine, *Annual Report 1988* (Washington, D.C.: National Academy Press, 1989), pp. 62, 81, 89–92.

23. IOM Council Meeting, Minutes, November 14–15, 1988, and July 24–26, 1989, IOM Records; "Food and Nutrition Board: Goals and Objectives," July 1989, Yordy Files, Accession 91-051, IOM Records; Institute of Medicine, *Annual Report 1988,* pp. 35–38, 62; Institute of Medicine, Catherine E. Woteki and Paul R. Thomas, eds., *Eat for Life: The Food and Nutrition Board's Guide to Reducing Your Risk of Chronic Disease* (Washington, D.C.: National Academy Press, 1992).

24. IOM Council Meeting, Minutes, July 21–22, 1988, IOM Records; Board on International Health Meeting, Minutes, January 17–18, 1989, Yordy Files, Accession 91-051, IOM Records.

25. *Ibid.*

26. Institute of Medicine, *Annual Report 1990* (Washington, D.C.: National Academy Press, 1991), p. 41.

27. Institute of Medicine, Christopher P. Howson, Polly F. Harrison, Dana Hotra, and Maureen Law, eds., *In Her Lifetime: Female Morbidity and Mortality in Sub-Saharan Africa* (Washington, D.C.: National Academy Press, 1996).

28. "New Directions for the Institute," June 1986, Yordy Files, Accession 91-045, IOM Records.

29. "Council on Health Care Technology: Chronology," in "The Council on Health Technology at the Institute of Medicine," April 1986; "Council on Health Care Technology, Interim Report, June 1987"; and Enriqueta Bond to J. Michael Fitzmaurice, Ph.D., Director, Department of Health and Human Services, October 2, 1989, all in Funding Files, Accession 91-045, IOM Records.

30. Glenn Kramon, "Guide Assesses New Technology," *New York Times*, May 17, 1988, p. D2; "Action Memo," October 22, 1988, Yordy Files, Accession 91-051, IOM Records.

31. IOM Council Meeting, Minutes, July 21–23, 1987, and April 20–21, 1987, IOM Records.

32. IOM Council Meeting, Minutes, November 14–15, 1988, IOM Records; "Access to Health Services," October 22, 1988, Yordy Files, Accession 91-051, IOM Records.

33. Institute of Medicine, *Annual Report 1989* (Washington, D.C.: National Academy Press, 1990), pp. 65–67.

34. IOM Council Meeting, Minutes, November 16–17, 1987, IOM Records; Samuel Thier to IOM Members, December 9, 1987, Yordy Files, Accession 91-051, IOM Records.

35. IOM Council Meeting, Minutes, January 9–10, 1989, IOM Records; Institute of Medicine, *Annual Report 1989*, p. 3.

36. IOM Council Meeting, Minutes, July 23–24, 1987, IOM Records.

37. IOM Council Meeting, Minutes, January 9–10, 1989, IOM Records.

38. Philip M. Boffey, "Experts Warn of 'Precarious' Vaccine Supplies," *New York Times*, July 30, 1985.

39. Samuel Thier to David Hamburg, December 5, 1986, Funding Files, Accession 91-045, IOM Records.

40. See, for example, Institute of Medicine, *Annual Report 1988,* p. 5.

41. Samuel Thier to Anne E. Hubbard, November 25, 1987, and "Proposal Draft," May 1987, both in Records of the Committee on Health Objectives for the Year 2000, Accession 90-067, IOM Records.

42. Testimony of Alvin Mauer, University of Tennessee, Osman Ahmed, Meharry Medical College, and William Hagens, Member of the Washington State House of Representatives, in transcripts of regional hearings, Birmingham, Alabama, January 14–15, 1988, and Seattle, Washington, February 5–6, 1988, in Records of the Committee on Health Objectives for the Year 2000, 1987–1990, Accession 90-067, IOM Records.

43. Mike Stoto to Committee on Health Objectives for the Year 2000, April 8, 1988, in Records of the Committee on Health Objectives for the Year 2000, Accession 90-067, IOM Records.

44. Mike Stoto to Committee on Health Objectives for the Year 2000, May 31, 1990, and Stoto to Committee on Health Objectives for the Year 2000, March 20, 1989, both in Records of the Committee on Health Objectives for the Year 2000, Accession 90-067, IOM Records; Institute of Medicine, *Healthy People 2000: Citizens Chart the Course* (Washington, D.C.: National Academy Press, 1990).

45. Institute of Medicine, *The Future of Public Health* (Washington, D.C.: National Academy Press, 1988), pp. 1–18.

46. Institute of Medicine, *The Future of Public Health: Summary and Recommendations* (Washington, D.C.: National Academy Press, 1988), p. 7.

47. On the importance of the report, see, for example, Institute of Medicine, *Annual Report 1988,* p. 4.

48. Institute of Medicine, Richard S. Dick and Elaine B. Steen, eds., *The Computer-Based Patient Record: An Essential Technology for Health Care* (Washington, D.C.: National Academy Press, 1991), p. 6.

49. *Ibid.,* p. 3.

50. Bruce Vladeck to Members of the Committee on Health Care for Homeless People, February 29, 1988, Homelessness Study Records, Accession 90-047-4, IOM Records.

51. IOM Council Meeting, Minutes, January 25–26, 1988, and July 21–22, 1988, IOM Records; Bruce Vladeck to Members of the Committee on Health Care for Homeless People, Homelessness Study Records, Accession 90-047-4, IOM Records.

52. IOM Council Meeting, Minutes, July 21–22, 1988, IOM Records; IOM *Annual Report 1988*, p. 5.

53. For background on this issue, see Edward Berkowitz, *Disabled Policy* (New York: Cambridge University Press, 1987).

54. Institute of Medicine, Marian Osterweis, Arthur Kleinman, and David Mechanic, eds., *Pain and Disability: Clinical, Behavioral, and Public Policy Perspectives* (Washington, D.C.: National Academy Press, 1987), pp. 2, 3, 8, 9; Edward Berkowitz and Richard Burkhauser, "A United States Perspective on Disability Programs," in Leo J.M. Aarts et al., eds., *Curing the Dutch Disease: An International Perspective on Disability Policy Reform* (Alderstot, U.K.: Avebury Press, 1996), pp. 71–91.

55. Institute of Medicine, *Annual Report, 1989*, p. 5; Jessica Townsend, Study Director, to David Nexon, Committee on Labor and Human Resources, October 10, 1989, and Townsend to Frank Press, November 8, 1988, both in Records of the Committee for Supporting Graduate Medical Training in Primary Care, Accession 91-007-5, IOM Records; Institute of Medicine, *Primary Care Physicians: Financing Their Graduate Medical Education in Ambulatory Settings* (Washington, D.C.: National Academy Press, 1989), pp. 10–11.

56. Relevant IOM reports included *Academic Geriatrics for the Year 2000* (1987); *Strengthening Training in Geriatrics for Physicians* (1993); *Extending Life, Enhancing Life: A National Research Agenda on Aging* (1991); *Allied Health Services: Avoiding Crises* (1988); *Role of the Primary Care Physician in Occupational and Environmental Medicine* (1988); and *Meeting Physician Needs for Medical Information on Occupational and Environmental Medicine* (1991), all published by the National Academy Press, Washington, D.C.

57. "Summary Report: Selected Cancers Study, Advisory Committee on the Centers for Disease Control, Study of the Health of Vietnam Veterans," April 25, 1990, Accession 90-068, IOM Records; Keith Scheider, "Agent Orange Study Was Obstructed," *New York Times*, August 12, 1990, p. A-12.

58. IOM Council Meeting, Minutes, July 21–22, 1986, IOM Records. In fairness, it should be pointed out that the IOM returned to this topic in its 1990 annual meeting.

59. IOM Council Meeting, Minutes, November 13–14, 1989, IOM Records.

60. IOM Council Meeting, Minutes, January 26, 1988, IOM Records.

61. IOM Council Meeting, Minutes, January 9–10, 1989, and April 10–11, 1989, IOM Records.

62. IOM Council Meeting, Minutes, November 13–14, 1989, IOM Records.

6

The Institute of Medicine and AIDS

"We are losing a generation of well-trained and exceptionally talented men," lamented June E. Osborn, dean of the School of Public Health at the University of Michigan, in a plaintive comment delivered in the course of what one observer described as a "remarkable" speech about the AIDS (acquired immune deficiency syndrome) epidemic at the Institute of Medicine's (IOM's) 1985 annual meeting.[1] She spoke at a discouraging time, when leadership in responding to the epidemic appeared to be lagging. President Ronald Reagan, for example, could not bring himself to utter the terms "AIDS" until the death of movie star Rock Hudson brought the subject into the open. Partly as a result of the silence of political leaders, the public appeared to be dangerously misinformed about the epidemic. At the same time that June Osborn, a student of Fred Robbins, gave her IOM speech on AIDS, a *New York Times*–CBS poll found that 47 percent of Americans believed that AIDS could be transmitted by sharing a drinking glass.[2]

In large part because of the IOM's activism on the AIDS issue beginning in 1985, the public began to receive better, more scientifically informed information on the subject, and the federal government made AIDS a higher priority in its research funding and public health activities. It would not be an exaggeration to say that the Institute of Medicine became the focal point of the nation's response to AIDS in 1986, and it continued to play a central leadership role in public policy toward AIDS in the years that followed. During the presidencies of Samuel Thier and Kenneth Shine, who became the head of the IOM in January 1992, the Institute issued two seminal reports on the nation's handling of the AIDS epidemic, conducted numerous workshops on nearly every facet of AIDS, and produced a series of reports on the government's management of the disease. In these ways, the Institute of Medicine realized its potential to serve as the voice of the scientific, public health, and medical communities in the face of a national emergency.

211

Facing Up to the Problem

In the summer of 1983, the IOM Council first took up the matter of AIDS at the urging of its members. One call to IOM President Fred Robbins came from Dr. Robert Ebert, the head of the Milbank Memorial Fund and later an influential figure in mobilizing the Sproull committee, who suggested that the IOM review the current research in the field. When Robbins raised this suggestion with Ed Brandt, the Assistant Secretary for Health at the Department of Health and Human Services (HHS), Brandt told him that there was little the IOM could do, even though Brandt had called AIDS the nation's number one priority in public health. Because Robbins had no desire to embarrass or upstage Brandt, who was an important supporter of the IOM within the Reagan administration, he assured Brandt that the IOM would do nothing until it received a specific request from the government. Council members requested that Robbins press Brandt on the matter and Robbins did so, only to receive the same response. The two agreed that the IOM would postpone any activity related to AIDS.[3]

This enforced passivity did not sit well with either Robbins or the IOM Council. In the spring of 1985, the IOM decided, on its own initiative and with Fred Robbins' strong endorsement, to dedicate its annual fall meeting to the subject of AIDS. By this time, many IOM Council members, such as John J. Burns, a chemist with a distinguished background as a researcher in the pharmaceutical industry, wanted the IOM to become more involved in public policy toward AIDS. Although Robbins sensed and shared the members' restiveness on this issue, he continued to believe that there was nothing the IOM could do "without the support and encouragement of the Public Health Service."[4]

Robbins did all he could to keep the pressure on the Public Health Service (PHS), and near the end of his presidency he sensed that the PHS attitude toward working with the IOM was gradually shifting. Sometime over the summer of 1985, he heard from James Mason, the director of the Centers for Disease Control (CDC), who suggested that the IOM examine the issues surrounding school admission policies for children with AIDS. At the time, the school attendance of children infected with what later became known as human immunodeficiency virus (HIV), the causative agent of AIDS, was a sensitive issue capable of igniting mass hysteria in an affected community. At the end of August 1985, the Public Health Service issued a set a recommendations on this matter in which the CDC stated flatly that "casual person-to-person contact as would occur among schoolchildren appears to pose no risk." Simply put, the CDC wanted company in making these sorts of pronouncements and sought out the IOM to add its authoritative voice and validate the CDC's conclusions. Robbins suggested that perhaps the time had come to pull together an IOM task force that would consider all aspects of the AIDS problem and identify the specific things that the IOM could do.[5]

On October 16, 1985, the IOM devoted its annual program to the subject of AIDS. Chaired by Philip Leder, the head of the Department of Genetics at the Harvard Medical School, and staffed by Enriqueta Bond, the meeting considered the scientific, ethical, sociological, and financial issues related to AIDS. The key speakers were government officials who reported on the scope of the epidemic and detailed the government's response. These included James Curran, chief of the AIDS branch of the CDC, Anthony Fauci, director of the National Institute of Allergy and Infectious Diseases at the National Institutes of Health (NIH); and Robert C. Gallo, chief of the National Cancer Institute's Laboratory of Tumor Cell Biology. The meeting generated such interest that with the aid of funds from Hoffman-La Roche and the National Research Council (NRC), and with the assistance of Bond and science writer Eve Nichols, the IOM transformed the talks into a coherent volume that Harvard University Press published in 1986.[6]

The volume served as a careful compendium of information related to AIDS. Written in accessible language, similar in tone to the IOM report summaries of the period, the book was intended to reach a wide audience of educated, but not scientifically sophisticated, readers. Here one could find a journalistic description of the epidemic's human face—"in cities across the United States, homosexuals describe the anguish of watching one friend after another weaken and die"—and an epidemiological description of the epidemic's course—"experts expect that a total of 14,000 to 15,000 new cases of AIDS will be diagnosed in 1986 and that at least 7,000 AIDS patients will die." The book also explained how scientists identified the virus that causes AIDS, unveiled the virus's genetic code, and produced an effective test for presence of the virus. Still, the volume could hardly be read as an ode to scientific progress. "These monumental achievements," Eve Nichols wrote, "do not mean that there will be a rapid solution to the AIDS problem."[7]

The problem for the nation and the world was how best to encourage a solution and what to do in the painful interim. On the day after its annual meeting, the IOM, with the encouragement of Frederick Robbins, held a follow-up workshop in which the invited participants tried to define an appropriate role for the IOM in combatting the AIDS epidemic. The discussion centered on applied science, such as the development of an AIDS vaccine or an antiviral drug. Less was said about the provision of care to AIDS patients and the many medical, social, and behavioral issues involved. Whatever their particular take on the problem, the participants agreed that AIDS was an appropriate subject for the IOM to consider.[8]

By the time the IOM Council next sat down to discuss this matter, Samuel Thier, rather than Fred Robbins, presided over the IOM. For at least two reasons, Thier wanted to finish what Robbins had begun. First, the subject of AIDS interested Thier, who had seen a large number of AIDS patients as chief of medicine at Yale. Second, Thier recognized that because the IOM's response to AIDS was largely unformed in November 1985, it would be amenable to his influence. AIDS had the potential to become one of the IOM's signature activities, a demonstration of the Institute's competence and newfound sense of activism.

In November 1985, the IOM Council unleashed a torrent of ideas about how the IOM should respond to AIDS. John Burns suggested that the IOM might coordinate trials for the antiviral drugs that were being developed to combat AIDS. Although it seemed unlikely that the IOM would usurp the role of the Food and Drug Administration (FDA), other activities appeared more plausible. Burton Weisbrod, an economist from the University of Wisconsin, said that the IOM should insert a calm and reasoned viewpoint into an often unreasonable debate because, as matters stood, too many people with too many irresponsible ideas threatened to dominate the discussion. If the IOM were to survey the field, Weisbrod argued, it would have to do so quickly, possibly completing its study in three months, because information moved too fast for the IOM to undertake one of its leisurely two-year studies. Gilbert Omenn, dean of the University of Washington's School of Public Health, countered that the subject was too important and too complex for such a cursory examination.

Listening to the conversation, Thier decided that "there was a strong sense among Council members that IOM should be involved in the AIDS issue." He announced his intention to meet with National Academy of Sciences (NAS) President Frank Press to discuss an Academy-wide, multifaceted approach to the issue. With such an approach, the IOM could speak to the government and scientific communities interested in vaccines and drugs, the public health community concerned with the provision of care, and the government and private insurers involved in financing the costs of care. Thier envisioned one overarching group to coordinate the process, with subgroups formed as necessary.[9]

Organizing an NAS response to the AIDS epidemic became Thier's top priority and the first project he tackled as IOM president. In less than three months, he persuaded Frank Press and the Council of the National Academy of Sciences to join in the effort, and he and Press convinced the National Research Council to support the project with its own funds. Thier and Press appointed an 11-person steering committee, cochaired by David Baltimore of the Whitehead Institute for Biomedical Research in Cambridge and Sheldon Wolff of the Tufts medical school, to develop "a national agenda on research and health care for AIDS." David Baltimore, a biologist from the Massachusetts Institute of Technology (MIT) who had won the Nobel Prize in 1975 for his work on retroviruses, headed the Research Panel. Among its members was Howard Temin of the University of Wisconsin, who had shared the 1975 Nobel Prize with Baltimore. Sheldon Wolff, the chair of the Department of Medicine at Tufts and an expert on infectious diseases, led the Health Care and Public Health Panel. It consisted of 10 people concerned in one way or another with public health, including June Osborn of the University of Michigan and David Fraser, the president of Swarthmore College.[10]

Toward the 1986 Report

These individuals were expected to produce a comprehensive report on AIDS in six months. On the scientific side, the National Academy of Sciences directed the committee to consider the appropriate actions necessary to develop, test, and distribute a vaccine and to develop "chemotherapeutic or chemoprophylactic agents." On the public health side, the NAS wanted the committee to provide advice on appropriate measures to protect the public from the disease and on the best means of treating the affected population and financing the treatment. The need to move so quickly was dictated by the size of the problem. At the time the Academy took action, public health authorities estimated that 20,000 Americans had full-blown AIDS, with few expected to survive for more than 18 months. The CDC hypothesized that between 1 million and 2 million Americans were HIV-positive, with at least 45 percent of these individuals expected to develop the disease. Rough estimates of the Medicaid costs of treating an AIDS patient in California over the course of the illness were about $59,000 per patient. Each of these statistics was an indication that "the spread of the disease has outdistanced public and private efforts to identify and coordinate the actions that are needed."

Hence, the Institute of Medicine and the National Academy of Sciences sought to play a leadership role on AIDS.[11]

Two things about the nature of the effort were noteworthy. First, the Institute of Medicine took the lead in developing the AIDS report and the National Academy of Sciences acquiesced. The report of the Committee on a National Strategy for AIDS appeared in October 1986 as a publication of the Institute of Medicine, rather than the National Academy of Sciences or the National Research Council. The IOM staffed the effort on behalf of the Academy, with Roy Widdus, staff director of the IOM's Division on International Health, detailed to the committee. The fact that the NAS allowed the IOM to direct such an important, high-profile activity of the Academy complex represented a vote of confidence in the IOM and its new president. It demonstrated that the stigma of the Sproull report was beginning to fade. Second, the IOM decided to initiate the project without waiting for an invitation from the CDC, NIH, HHS, or some other component of the federal government. Contrary to the usual rhythms in health policy, the Institute of Medicine moved more quickly than did the federal government. The Committee on AIDS resembled nothing so much as a presidential commission, yet it was the IOM, not the federal government, that did the commissioning. In effect, the IOM filled a void left by the government. In the past, the IOM had often evaluated actions taken by the federal government. In the AIDS effort, the IOM acted in advance of the federal government.

The committee's efforts got under way at the beginning of March 1986, with a special workshop on AIDS epidemiology and disease burden projections that was intended to provide the committee with a good road map on the size and shape of the problem. J. Thomas Grayston, a professor of epidemiology at the University of Washington, who headed the epidemiology working group advising the committee, chaired the meeting, which showed just how difficult a topic AIDS was. Part of the problem was the lack of reliable information that came from the news media. As James Curran, the medical epidemiologist who worked on AIDS for the CDC, stated, "A lot of the stuff is a bunch of garbage that you read in the newspapers." Another part of the problem was the fact that no one had a good handle on the potential size of the epidemic. "How many people are infected?" Curran asked. "The answer is, and this is open to challenge, no one knows." As for the ultimate case fatality rate, Curran believed that "the ultimate case fatality will be higher than 10 percent. Whether it is 50 percent or 100 percent or 40 percent is unknown, in my opinion."[12]

Against this backdrop of uncertainty, the steering committee and the two panels met for the first time on March 13, 1986. The first meeting produced less posturing and turf protecting than usual at the start of an IOM project, in part because all of the committee and panel members understood the gravity and urgency of the task and in part because the committee and panels covered such a wide range of professional backgrounds and expertise. The general attitude was one of wanting to get up to speed on the topic of AIDS as quickly as possible.

Even as committee members took a crash course on the AIDS epidemic, they grappled with the practical problems that inhibited action in the field. They decided, for example, to conduct two public hearings, one on either coast, so as to counteract the charge, in David Baltimore's words, "of being insensitive to the perspective and needs of the communities which are now shouldering the burden." The community of male homosexuals was chief among these, and as the committee learned, the fact that the syndrome disproportionately affected an already stigmatized group greatly complicated the politics of public health. Jerome Groopman, a hematologist at the Harvard Medical School, told, for example, of hearing rumors that money to educate members of the gay community on safe sex practices was not distributed because of a belief that such effort "directly discussed gay sexual practices" and hence "essentially condoned sodomy." The risk of reinforcing socially unacceptable practices applied even more to intravenous (IV) drug users because lessening the transmission of AIDS for this community meant making it safer to take illegal drugs. As one committee member expressed this idea, "At what point does education for the risk group be seen by some group as license or teaching people how to be addicts effectively and how to be gay?"[13]

Complicating the situation, the public appeared to favor repression over license and denial over action. At first, public health officials could not get anyone interested in AIDS. Walter Dowdle of the Public Health Service told the committee that 1981 and 1982 were "very discouraging years" in which it was difficult to get local communities to believe "it was something that we did have to pay attention to and indeed this was a problem that was going to affect all of us." Once past the smug reassurance that AIDS was something that happened to someone else and would soon succumb to the march of modern science, people reacted with hysteria. Again ignoring the best advice of public health experts, state legislatures considered bills calling for suspending HIV-positive individuals from handling food or prohibiting HIV-positive children from attending school. Paul Volberding of San Francisco General Hospital, on the front lines of the nation's fight against AIDS, said that "people enjoy the thrill of the fear of AIDS and don't want to hear the reassuring things."[14]

Not only was AIDS a public health challenge, it also posed considerable problems for the scientists on the committee's Research Panel. As Irving Weissman of Stanford Medical School put it, "This virus stretches our knowledge." As a result, producing a vaccine to prevent AIDS looked to be a complicated matter. In the near future, at least, it did not appear as though the AIDS crisis would end, as the polio crisis did, with the introduction of a new vaccine. "I think it is very important," said Howard Temin, "that the report not be able to read that a vaccine is soon going to come." Although David Baltimore agreed, he thought it important for the committee to give some sort of timetable for the development of a vaccine, such as 10 years.

Most of the scientists accepted the fact that a great deal of useful work had been done on the AIDS virus and agreed on the need for further work. Where they disagreed was on how to structure research activities. Maurice Hilleman of the Merck Institute for Therapeutic Research called for "a coordinated national effort" involving both the public and the private sectors. For his part, Howard Temin worried about a top-down, "centralized" campaign. He preferred to have "lots of different groups, very well supported, working on the common problem." In other words, Hilleman saw the fight to eradicate AIDS as something like the Manhattan Project, in which scientists and engineers banded together under central direction and produced the bomb. Temin believed that the attempt to halt AIDS should be science as usual, only more so: more investigators, more experiments working not in concert but on a common intellectual problem. Although most of the scientists did not disagree with Temin, they still felt the need to create "some kind of group over and above the battle to take a look at it" and to make sure that all appropriate resources had been brought to bear on the problem. In this spirit, David Baltimore closed the first meeting of the Research Panel on a fervent note. "What I am interested in," he said, ". . . is the number of people who accept immediately the fact that we are in a situation of national need, maybe not in the form of a Manhattan Project, but certainly on an order of magnitude that we have not seen in our lifetime."[15]

The message from the scientists was that a "cure" for AIDS would take a long time to develop, even under the warlike conditions that scientists hoped to create. This made it all the more important for the health care system to learn to treat patients with AIDS. In this area as in others, the committee discovered that AIDS confounded conventional wisdom. Because AIDS patients developed so many problems in so many different parts of their bodies, their treatment threatened the "subspecialty approach to medicine." As one doctor from San Francisco told the committee's Health Care and Public Health Panel, AIDS "is a much more complex disease than most others that we've faced. . . . All areas of medical care end up being involved in these patients." Diseases that affected the general population in one way affected AIDS patients in another, upsetting traditional treatment regimens. For a health financing system coming to depend on "diagnosis-related groups" (DRGs) that matched particular diseases or procedures with a pattern of hospital reimbursement, AIDS posed a considerable challenge. One official told the committee that "if you try to make one DRG out of AIDS, it would be one of the worst DRGs in terms of homogeneity, in terms of length of stay within that DRG that we've seen."[16]

The first meeting of the committee ended with only a general agreement that the nation needed to do more to respond to AIDS. The chief points of disagreement continued to be over how such an effort should be organized. The disagreements, in turn, reflected different readings of recent history. For some, such as David Baltimore, the war on cancer had been a great success, and as one participant put it, "the thing that made the difference . . . was simply dangling money in front of people." For others, such as Sheldon Wolff, the war on cancer was a "total flop." The difference in opinion was also one of perspective. For the scientists, the war on cancer was a success because it had led to great deal of creative science. For the clinicians, the war on cancer was a more limited success because it had not put an end to the disease. Similar considerations applied to AIDS. The nation would have to decide how much to spend on the promise of cure through basic science and how much to spend on care through public health and clinical measures. In the mid-1980s, unlike the early 1970s when the war on cancer was declared, the choices were more stark. "We live under Gramm–Rudman," said Sheldon Wolff, referring to an agreement in place to limit the rate of growth in the federal budget, "it is not business as usual." David Baltimore regarded AIDS as the leading health problem in the country, requiring the diversion of funds from other priorities. Sheldon Wolff believed that AIDS was a very serious problem but that others were also serious. Hence, the committee should be cautious about declaring war on AIDS if this meant diverting funds from other areas of health policy, particularly because funds were so tight.[17]

The committee continued to debate these issues through the spring and into the summer of 1986. Often the committee heard from AIDS activists in public and from leading scientists and public health researchers in private. The public hearings took on the aspects of a ritual, in which a person representing a particular group, such as the National Association of Gay and Lesbian Professionals, made a short statement for the record. The committee members who attended the hearings in New York and San Francisco almost never asked questions of the presenters. By way of contrast, they had many questions for the experts, such as health statistician Dorothy Rice, who briefed them on various aspects of the problem.[18]

Inside the Research Panel, scientists debated, as they had from the very beginning of the effort, whether to recommend some sort of central committee to coordinate and evaluate AIDS activities. In arguing for a committee, Baltimore pointed to such bureaucratic failures as the lack of close collaboration among the institutes of NIH, the failure of the Department of Defense to take advantage of resources at the CDC and NIH, the absence of contact and the presence of competition between the CDC and NIH, and the lack of interchange between NIH and the external scientific community. Each of these factors indicated that the government was not responding as effectively to the AIDS crisis as it might. Baltimore's remedy was an "ongoing commission" to monitor the course of the disease, evaluate research efforts, and catalyze coordinated efforts. The panel both accepted and tempered this argument. It did not favor a commission that had the power to allocate funds to particular researchers or research establishments. Instead, it decided on an advisory commission, perhaps operated by the Institute of Medicine, that would operate through moral suasion and public relations and would be independent of the federal government. In fact, the panel recommended a commission much like itself.[19]

The 1986 Report

In the final report, issued on October 29, 1986, Baltimore's recommendation of an ongoing commission took the form of a National Commission on AIDS appointed not by the Institute of Medicine but by the President of the United States or jointly by the President and Congress. It was the committee's way of indicating that AIDS required the attention of the nation's political leaders, not just the medical and scientific communities. In general, the committee tried to bring a sense of urgency to the problem. On the public health side, it called for a "massive media, educational and public health campaign to curb the spread of the HIV infection." On the scientific side, it advocated "substantial, long-term and comprehensive programs of research in the biomedical and social sciences intended to prevent HIV infection and to treat the diseases caused by it." The committee advised that these programs would cost $2 billion annually by the end of the decade, all part of "the most wide-ranging and intensive efforts ever made against an infectious disease."[20]

Beyond these major recommendations, the committee also addressed many other aspects of the AIDS crisis. To curb the spread of HIV infection, the committee suggested that more people receive serologic testing, although it favored voluntary testing with the results kept confidential. The committee also cautiously recommended that clean needles and syringes be more freely available to IV drug users in order to reduce the sharing of contaminated equipment. To treat AIDS patients, the committee called for more planning and training so as to cope with an increasing case load of patients with HIV infection. The committee sought to emphasize care in the community, keeping hospitalization to a bare minimum. To further the research effort, the committee cited the need for basic research in virology, immunology, and viral protein structure, as well as the need for more epidemiological research to trace the spread of the HIV infection. Academic scientists could be encouraged to participate in research against AIDS through an increase in funding for investigator-initiated research proposals, but the federal government should also solicit the participation of private industry in this effort. A key part of the research infrastructure that required attention was an expansion of "experimental animal resources," with careful conservation of the chimpanzees that were used in AIDS experiments. The United States should also recognize the international dimensions of the AIDS crisis and become a full participant in international efforts against the epidemic, including support of World Health Organization (WHO) programs.[21]

Not content to stop there, the committee also addressed questions related to civil rights, making the flat statement that "discriminating against those with AIDS or HIV infection because of any health risk they may pose to others in the workplace or in housing is *not* justified and should *not* be tolerated." Further, the committee worried about discrimination against individuals simply because they were members of high-risk groups and recommended that "any form of discrimination against groups at high risk for AIDS should be prohibited by state legislation and, where appropriate, by federal laws and regulations."[22]

Viewed as a coherent document, rather than a laundry list of recommendations, the report reflected an interesting blend of scientific, medical, and sociological concerns. On the one hand, the report praised federal research efforts. On the other, the report conceded that the development of an AIDS vaccine was still at least five years away. This meant that the nation should both support scientific research that would pay long-term dividends and invest in an educational campaign that represented the nation's best short-term hope. The educational campaign, according to the report, deserved just as much money as the scientific campaign, a pivotal concession from a group with NAS sponsorship and dominated by scientists. As for the terms of the education campaign, the committee recommended that it be targeted at high-risk groups, including homosexual men, intravenous drug users, sexually active heterosexuals, and teenagers. It was a time for plain speaking. The messages should be as direct and as explicit as possible, using whatever language people could understand and reflecting the basic reality that anal or vaginal intercourse with an infected person without using a condom was risky behavior.[23]

Nothing in the entire history of the Institute of Medicine attracted as much interest as the 1986 report on AIDS. The report received saturation coverage in the newspapers, including page one stories in both the *New York Times* and the *Washington Post*. The *Times* chose to highlight the political contest aspect of the story. Reporter Philip Boffey's piece led with the accusation that the federal response to AIDS was "dangerously inadequate." The third paragraph contained a quotation from David Baltimore about how he was "quite honestly frightened" by the potential of the AIDS epidemic, and the sixth paragraph featured a response from Robert Windom, an Assistant Secretary of Health and Human Services, to the effect that "there has been a conscientious effort to do a good job." The lead paragraph identified the report as from the National Academy of Sciences, whose arguments carried weight "because of its reputation and function." Not until the eighth paragraph, which was buried inside the newspaper, did the interested reader learn that the committee had been assembled "by the Academy and its principal health unit, the Institute of Medicine." The price of collaboration with the NAS was subordination.[24]

The *Post* story, by way of contrast, mentioned the Institute of Medicine in the second paragraph and on equal terms with the National Academy of Sciences. Christine Russell's piece also highlighted the public health aspects of the report more than the *Times*. She quoted Sheldon Wolff as saying that "people should be told that they can protect themselves against the disease by using condoms during sexual experience—either anal or vaginal—with an infected or possibly infected person and by not sharing needles and syringes." Only toward the end of the story did Russell indulge in political speculation, referring to some committee members' private complaints that "the White House has been very quiet on this issue" and giving Congress much of the credit for federal leadership.[25]

Although such nuances mattered to Washington insiders, most people received their news from television, not the newspapers. It was therefore highly significant that all three networks chose to lead their evening broadcasts on October 29 with the AIDS report. On ABC, the audience saw a text box that highlighted the report's major recommendations and heard interviews with Samuel Thier, health panel member James Chin, and David Baltimore. The network devoted nearly four minutes to the report, a substantial investment of network time. Dan Rather of CBS gave the report two and a half minutes, including interviews with Thier and Baltimore. Tom Brokaw and NBC spent more than four minutes on the report and ran a story, reported by Robert Bazell, that featured interviews with Baltimore, Chin, and Sheldon Wolff. The NBC coverage featured "cross-talk" between anchor Brokaw and correspondent Bazell on the subjects of human testing of vaccines and funding for research.[26]

Of all the participants on the AIDS committee, David Baltimore and Sheldon Wolff played the key roles, both in the composition of the report and in the public relations blitz that followed its release. "We should be decorating you as a combat veteran of the first 'fast track' study of the Institute of Medicine/National Academy of Sciences," Thier and Frank Press wrote to the pair. "Certainly you have noticed and contributed handsomely to the volume of news media coverage of the report, which must have established some high-interest mark for one of our publications."[27]

The 1986 AIDS report put the IOM at the center of the health policy universe, at least for a day. Not since the heart transplant statement issued in February 1968 had the IOM come anywhere near this level of influence. The contrast between the heart transplant statement and the AIDS report indicated some of the differences that had taken place since the founding of the IOM. In 1968, the Board on Medicine attempted to slow down the diffusion of heart transplant surgery. It saw the matter as one in which medical knowledge outstripped its application to human needs. In 1986, the Institute of Medicine sought to speed up the rate of research into the problems of AIDS. It saw the problem as one in which medical and scientific knowledge lagged far behind human needs. In 1968 the problems of medicine appeared to be those of controlling abundance: too many people in too many hospitals wanted to perform heart transplants. In 1986 the problems of medicine appeared to be those of coping with scarcity: too few people in too few hospitals wanted to care for AIDS patients. The differences between the two situations illustrated how history could confound expectations. Few people could have predicted in 1968 that the IOM would gain its greatest fame for its efforts to come to grips with an infectious disease for which there was no apparent cure or clear medical treatment.

AIDS Workshops

Having gotten the nation's attention, the IOM had to decide what to do with it. In November 1986, the IOM Council agreed, with no hesitation, that it wanted the IOM to continue its efforts in the area of AIDS. The consensus was for the IOM to form a small group that would monitor the implementation of the committee's recommendations and help facilitate international activities in the AIDS field and activities involving private pharmaceutical companies. In the spring of 1987, Thier and Press once again joined together to form an AIDS oversight committee.

In the fall of 1986 and the first months of 1987, the IOM waited to see how the Reagan administration and Congress would react to the 1986 report. At the end of June 1987, President Ronald Reagan announced that he would create a Presidential Commission on the Human Immunodeficiency Virus Epidemic. When the IOM Council learned in July who would be named to this commission, it expressed a collective sense of dismay. As Thier delicately expressed it, "The number of individuals on the commission with in-depth knowledge of AIDS was very modest." The Council decided nonetheless to adopt a "respectful and communicative attitude" toward the commission, "without being co-opted."

The presence of the Presidential Commission on the HIV Epidemic spurred the IOM to move ahead with its own AIDS oversight committee. The primary mission of this committee would be to review and update the 1986 report. In addition, the IOM hoped that the oversight committee would help focus the AIDS activities of the entire NAS complex, serving in a similar manner to the boards that guided the various IOM divisions.[28]

Even before the oversight board started its work, the IOM initiated limited projects in the areas of epidemiology and AIDS drug and vaccine development. At the end of August 1987, the IOM held a Conference on Promoting Drug Development Against AIDS and HIV Infection. The conference followed directly from a recommendation in the 1986 report to "convene researchers from industry, academia, and the government to improve the development of drugs against AIDS and the virus infection that causes it." As such, it provided an opportunity for basic scientists, federal bureaucrats, representatives from drug companies, and AIDS activists to discuss issues in the field. The group considered the state of science in AIDS drug development, current federal and industrial activities, issues in the clinical evaluation of drugs, appropriate levels of access to experimental drugs by AIDS patients, and whether or not there should be a "national strategy" in the effort to develop a drug treatment for AIDS. "I believe we are in a national emergency and the biomedical community will be judged in the future by how we respond," David Baltimore said in a keynote address that set the tone for the conference. David W. Barry, vice president of research at Burroughs Wellcome, FDA Commissioner Frank Young, Sheldon Wolff, and former Congressman Paul Rogers also spoke. Rogers, for one, was pleased with the outcome. An unfailingly courtly and courteous man, he wrote Sam Thier that the conference was "most successful. . . . I think this format is certainly helpful to the academic, industrial, and governmental sectors. Getting people to talk out their problems and what needs to be done is certainly an effective way for the Institute of Medicine to operate."[29]

This was the main point of the conference: getting people to talk out their problems in a manner similar to the IOM's Forum on Drug Development and Regulation. The conference was intended to capture knowledge that could be shared easily among the loosely organized partners in the national AIDS drug development effort. Although Roy Widdus of the IOM's Division of International Health did not expect the conference to yield a product with a long shelf life, he and his staff prepared a report that became an IOM publication. In the traditional manner of IOM reports, it included a list of recommendations, the chief of which was that "systematic screening of compounds in the libraries of the world's pharmaceutical companies offers the best short term prospects and should be vigorously pursued." The report recommended that scientists, in effect, continue to "grind out" results. In the meantime, the public needed to receive as much information as possible, both to guard against false hopes and to keep people with AIDS or HIV infection away from the "expanding underground sources of potentially toxic agents." The drug area remained a particularly important part of the battle against AIDS, because the development of a vaccine appeared to be a long way off. The 1986 report had estimated a minimum of five years to produce a vaccine. "Things today seem, if anything, bleaker," David Baltimore told the conference participants.[30]

The 1986 report had called attention to the international aspects of AIDS and recommended that the United States be "a full participant in international efforts against AIDS and HIV infection." As a step toward facilitating this involvement, the IOM organized a workshop, held in October 1987, that focused on the epidemiology of AIDS in an international context. The United States Agency for International Development funded the small workshop that was intended to review "the art of modeling of HIV transmission and the demographic impact of AIDS." Burton Singer, head of Yale's Department of Epidemiology and Public Health, chaired the steering committee, which also included authorities from Europe, Canada, and Africa. Participants at the workshop came from around the world, but all spoke a common statistical language. Despite the elegance of their methods, the epidemiologists and demographers admitted that the "development of reliable, specific long-range predictions was out of reach of current or foreseeable model capabilities and data." The data did suggest that the number of AIDS cases "was certain to grow for many more years" and, in particular, that AIDS in Central Africa would cause "substantially increased mortality in the general population." Participants agreed that they needed to know much more about sexual behavior—the frequency of sexual contacts, the duration of partnerships, the selection of partners—in order to understand the course of the epidemic. They realized, however, that the "ethics of answering questions about sexual behavior" differed from country to country, complicating the process of creating an international data set. In the end, the workshop on modeling AIDS, like the conference on drug development, highlighted problems rather solutions. It also served as a forum for a much-needed international discussion on the size and shape of the AIDS epidemic.[31]

At the end of 1987, the IOM held a third workshop on AIDS, this one concerned with vaccine development. Support came from the Pharmaceutical Manufacturers Association, the National Cancer Institute, and the National Institute of Allergy and Infectious Diseases. David Baltimore headed the small steering committee for this conference, which included experts in immunology, retrovirology, vaccine design, and vaccine evaluation. Baltimore began the conference with another gloomy assessment about the prospect of a vaccine. He vowed, however, to continue the work "along every possible line of attack because the only way that science can produce its surprises is if we keep working at it, but I suspect that in the long run we are going to need more than a standard surprise." The conference was intended to be a meeting of scientists talking to other scientists, with the focus on the science itself rather than on public policy issues associated with the management or conduct of science.[32]

Because data on AIDS changed so quickly, the conference report was less valuable than the conference itself, even though the steering committee did prepare a short summary with some preliminary conclusions. Much of this summary contained what to nonscientists was an impenetrable description of HIV infection. Part, however, focused on the more accessible, and very important, issue of the conduct of clinical trials. A majority of conference participants felt that tests on humans should not be done until there was evidence of "protective efficacy from animal tests." The steering committee recommended that a vaccine should proceed to clinical trials only after it had been shown to be effective in chimpanzees, macaques, or some other suitable laboratory animal, unless the vaccine's design was based "on a fundamental new understanding of relevant human immune responses that cannot be adequately modeled on animals." In all likelihood, such clinical trials lay far in the future. "Experimental evidence has not yet justified any hope for an effective vaccine in the next few years," the steering committee concluded.[33]

Toward the 1988 Report

By the time the conference on vaccine development took place in December 1987, the AIDS Oversight Committee had begun to meet. Samuel Thier selected Theodore Cooper to head the eight-person committee. An M.D.-Ph.D. with a background in cardiovascular surgery and a major figure at the IOM for a long time, Cooper had served as director of the National Heart and Lung Institute in the late 1960s and 1970s, before becoming an Assistant Secretary for Health in the Ford administration and then the provost for health affairs at Cornell. In the fall of 1987, Cooper held the positions of vice president and vice chairman of the board of the Upjohn pharmaceutical company. Joining Cooper on the committee were David Baltimore, Howard Temin, political scientist and Rockefeller Foundation official Kenneth Prewitt, Oregon State Health Division administrator Kristine Gebbie, international health expert Donald Hopkins, AIDS health care authority Paul Volberding, and Stuart Altman, an economist with expertise in health financing. With the exception of Baltimore and Temin, who respected but often disagreed with one another, each committee member had a well-demarcated area of expertise that gave him or her a distinctive niche and enabled committee members to work effectively with one another.[34]

From the beginning, the Oversight Committee realized that it had to respond to two different audiences. One was an external audience of members of the Washington political and health policy communities; the other was an internal audience of National Academy of Sciences officials. To reach the political audience, the Oversight Committee hoped to issue an update of *Confronting AIDS* in time to influence the 1988 party platforms. After the election, the committee expected to present a short version of the report to the President-elect. The internal NAS audience required not policy recommendations so much as advice on the coordination of the Academy's AIDS studies. In this capacity, the Oversight Committee would evaluate project proposals and try to keep one branch of the NAS aware of what the other branches were doing. As a practical matter, preparation of the update took precedence, because the committee, which met for the first time in October 1987, faced a short deadline for its completion.[35]

Because Samuel Thier regarded the work of the AIDS Oversight Committee as having the highest priority of any IOM activity, it frustrated him that funds for the committee and for other AIDS projects were difficult to raise. In November 1987, he told the Council that money for the report update was "being pieced together," with the bulk of the support coming from the National Research Council. Part of the problem was the presence of the Presidential Commission on the HIV Epidemic, whose work was being done simultaneously with that of the Oversight Committee. It was natural for funders to wait until the commission reported before supporting another such group. Hoping to get money from the Department of Health and Human Services, Thier told Assistant Secretary for Health Robert Windom that a natural division of labor existed between the two groups, in which the IOM concentrated on science and the commission on other public policy matters. In fact, however, Thier held the commission in low esteem and expected little of it. When the commission requested IOM assistance, Thier replied that he wished to be as helpful as possible, but in the privacy of the IOM Council, Thier said that he did not want the IOM identified with the commission and would not subcontract with it. He told Leighton Cluff, president of the Robert Wood Johnson (RWJ) Foundation that the "problems of the President's Commission," in particular the fact that many scientific and medical problems "appear beyond their capacity to address," only underscored the need for the IOM committee. "Given . . . the difficulties the commission has been experiencing, it seems more important than ever we proceed . . . and expand our efforts," he wrote.

Although Cluff might have agreed with Thier's take on the situation, he still turned down the request for funds. He argued that RWJ had already given the IOM a large grant for its endowment and this precluded the foundation from making specific project grants.[36] In order to publish and disseminate the update report, the IOM spent $50,000 of its own money, a good example of how the drive to raise an endowment enabled the IOM to accomplish things it might not otherwise have been able to do.[37]

In the spring of 1988, the AIDS Oversight Committee came to an agreement on the recommendations for its update of the 1986 report. In general, the committee used the 1986 recommendations as starting points and then looked at what had occurred in the interim that might cause them to be modified or changed. Despite the presence of the Presidential Commission, the committee remained convinced of the need for a "permanent AIDS commission" with both congressional and administration representation. Within the committee, Cooper argued for this recommendation on the basis of the fact that the Presidential Commission had been ineffective and that there had been a lack of leadership on the part of the Reagan administration. He had in mind a commission similar to the one that had investigated Social Security financing and helped to produce an agreement leading to legislation in 1983. This group had been successful because it contained administration and congressional representation. When Paul Volberding complained that the Reagan administration's conservative nature made it unable to deal with the AIDS epidemic, Cooper replied that this was exactly why Congress had to be involved. Thier concurred in this decision and did not seek to carve out an expanded role for the IOM in any commission that might be formed, believing that the Congress and the President were the responsible political actors. In a similar manner, the group came to agree that scientists, like the physicians in the IOM, should not manage the nation's response to AIDS; rather they should be insulated from political pressure and left alone to do their work.[38]

In reviewing the nation's response to the AIDS epidemic, the committee discovered that reducing AIDS transmission among IV drug abusers constituted "the most serious deficiency in current efforts to control AIDS in the United States." Donald Hopkins, who took the lead for the committee in this area, said that prevention and treatment of IV drug abuse was a "big hole."[39] The committee considered both long-term and short-term strategies for dealing with the problem. In the short run, Hopkins proposed "rapid expansion of drug abuse treatment slots" and "immediate extension of serologic testing and counseling." In the long run, he suggested efforts to dissuade teens and others at high risk from taking drugs and an evaluation of the efficacy of providing users with sterile needles.

Discrimination was one area in which the committee felt the need for action, not evaluation. Since the 1986 report, even more evidence had arisen to show that discriminating against people with AIDS or HIV because of the health risks they posed to others in the workplace or housing was not justified. Hence, legal measures to prohibit discrimination against people with AIDS remained "a vital part of public health efforts."[40]

Health care financing marked another area in which the AIDS Oversight Committee sought to expand the work of the original NAS–IOM committee. Stuart Altman developed a proposal that the committee debated and ultimately endorsed. He pointed out that the health care financing system was pluralistic and unlikely to be centralized in the near future and that most people were concerned that the nation was spending too much, not too little, on health care. Furthermore, AIDS patients were themselves a diverse group and the severity of their medical problems varied widely. He recommended a categorical grant program under which federal money would be distributed to localities for the treatment of people with AIDS or HIV under a managed care system. Kristine Gebbie wondered if AIDS should be a special case or whether the nation should instead fix the entire health care financing system. The committee decided that AIDS should indeed be a special case.[41]

The 1988 Report

On June 1, 1988, the committee issued *Confronting AIDS: Update 1988,* which argued that despite progress in biomedical research and public education, inadequacies—in particular a failure to combat AIDS related to drug use—remained in the nation's response to AIDS. The committee urged a multipronged approach that included the expansion of research facilities, the targeting of public education programs on the groups most at risk, the expansion of drug treatment programs, and the health care financing plan that Altman had outlined. The 1988 report made much stronger statements on the relationship between HIV and AIDS than the 1986 report. The 1988 group recommended that "HIV infection itself should be considered as a disease" and cited the view of analysts who believed "that virtually all HIV-infected individuals will eventually develop AIDS." It also pointed to the efficacy of education efforts, such as those of U.S. Surgeon General C. Everett Koop, and urged their expansion, "including the use of paid advertising on television," rather than bland, and often unwatched, public service announcements. As for research, the report stressed the points made by David Baltimore that "we are no closer to having a licensed vaccine against HIV than we were two years ago" and that drug development offered "the best hope of slowing the epidemic through research."[42]

Release of the 1988 report brought another avalanche of publicity to the IOM. In this case, however, coverage was tempered by the fact that the administration no longer presented such a large target. For one thing, Surgeon General Koop had emerged as a public health hero, widely praised for his advocacy of safe sex practices. For another, the Presidential Commission on the Human Immunodeficiency Virus Epidemic, once vilified as ignorant and stacked with what *Science* described as "homophobic right-wing ideologues," appeared on the verge of issuing a report with recommendations that mirrored those of the IOM. In fact, the IOM report used cost data from the Presidential Commission to bolster its call for increased spending on public health and educational efforts. Agreement was always less of a story than disagreement. Finally, and most importantly, the presidential campaign was well under way in June 1988, and whatever the outcome, Ronald Reagan would no longer be President in 1989. Although the *Times* ran the story of the AIDS update report on page one and reprinted excerpts from the report, it mentioned that the report was only the "opening salvo in an unusually busy week for appraising the AIDS epidemic" and referred to the fact that James D. Watkins, chairman of the Presidential Commission, would issue its final conclusions that day. The *Times* also reported that the IOM's recommendations to increase drug treatment facilities added weight "to similar recommendations by the President's AIDS Commission." Instead of writing a story on the IOM report, *Science* chose to highlight the consensus between the two reports.[43]

Even if the IOM could no longer lay exclusive claim to the field, its two reports served as important catalysts for action. As the *Times* noted in its story on the 1988 report, the 1986 report had "provided a benchmark by which many members of Congress and analysts judged the effectiveness of the nation's effort to combat AIDS." No doubt, the report helped to spur legislation that nearly doubled federal spending against AIDS from fiscal year 1987 to fiscal year 1988 and motivated the payment of U.S. dues for WHO programs on AIDS. According to the IOM, the 1988 report led to amendments to the Public Health Service Act that authorized $1.5 billion for research and public health measures related to AIDS and permitted the CDC to develop AIDS-related advertisements and put them on the air.[44]

Advice for the President

For the Oversight Committee, preparing a white paper on AIDS for the next administration became an immediate priority after issuing the update. Earlier in the year, the presidents of the National Academy of Sciences, the National Academy of Engineering, and the Institute of Medicine had decided to produce a series of short issue-oriented papers on critical issues in science and technology, with the incoming administration as the intended audience. AIDS emerged as an obvious choice for such a paper, with the 1988 update report serving as the basic source of data. It fell to Robin Weiss, director for AIDS activities at the Institute of Medicine, to summarize the update report in a way that highlighted actions George Bush and his administration should take to stem the epidemic.[45]

In the interim between the update report and the white paper, an important piece of legislation, the AIDS Amendments of 1988, changed the political outlook on AIDS. This legislation followed the advice of the Institute of Medicine and established a National Commission on Acquired Immune Deficiency Syndrome. Under the term of the law, the President would appoint two members to the commission and would in addition ask the Secretary of Health and Human Services, the Administrator of the Department of Veterans Affairs, and the Secretary of Defense to serve *ex officio* as nonvoting members. In the white paper, the IOM urged the President-elect to choose "senior experts" on AIDS, without regard to political ideology. Congress was to play the dominant role in the appointment of voting members, with five Senate and five House appointees to be chosen.[46]

The white paper had the compressed tone of a document written to attract the attention of a busy person. Aware that the President-elect would not have time for a leisurely examination of even a 13-page document when he received it in December 1988, Frank Press and Samuel Thier wrote a 3-page cover letter in which they summarized the white paper. They began with the stark fact that during George Bush's term in office, AIDS would kill 200,000 Americans. Another 1.3 million were infected with HIV, and most would die. Press and Thier urged Bush to "use the newly legislated national commission on AIDS to develop a forceful and coherent national policy," to encourage antidiscrimination legislation, to express support for an "aggressive, unambiguous education program," and to recognize America's "special responsibility in international efforts to control the spread of HIV infection." "Your Administration inherits the opportunity to harness our knowledge and turn the tide against AIDS," wrote Press and Thier.[47]

Having invested time and effort in creating the white paper and composing the cover letter, Thier set out to get Bush to read it. At Thier's urging, Monroe Trout, chairman of the board at American Healthcare Systems, wrote to Barbara Bush asking her help "to get a paper read by George." He advised Mrs. Bush that she might want to read the paper herself, "even though it is not the most pleasant of subjects." Trout then told Thier that he had sent off the material to the Vice President's office and "hopefully he will read it." How Bush felt about the white paper or whether or not he read it remains unclear. Thier and Press did receive a letter from him thanking them for the white paper and containing some general remarks about AIDS. "The work of the National Academy of Sciences and the Institute of Medicine has strengthened my determination to see to it that out of this international tragedy some good may come; and that America's unparalleled medical resources [will] be stronger for having fought this fight," wrote Bush.[48]

The Institute's AIDS Activities

After the excitement of preparing the update, the AIDS Oversight Committee settled into the role of advising the IOM on its AIDS activities. One of the first projects scrutinized by the committee was a proposed IOM evaluation of the NIH AIDS programs. David Baltimore regarded the evaluation study as a way of continuing the IOM's leadership in the field. It would offer an opportunity to evaluate the NIH response to the epidemic, to suggest an "ideal" research program, and to offer insights on how NIH should respond to future public health threats. A good study could also examine how money to fight AIDS was divided among the various institutes, explain why NIH responded so slowly to AIDS, and determine how NIH coordinated its activities with other parts of the Department of Health and Human Services. The Oversight Committee even suggested names for the steering committee and created a detailed plan of action for the study, complete with a list of people who should testify at a public hearing.[49]

For all of the hopes of the AIDS Oversight Committee about the many things the study might accomplish, it was similar to many other evaluations of the government's research programs that the IOM had undertaken over the years, and it was likely to produce the same sort of recommendations. In this case, the IOM responded in 1988 to a specific request from James B. Wyngaarden, the director of the National Institutes of Health, and Anthony S. Fauci, the associate director for AIDS research, to consider such questions as the effectiveness of NIH's use of advisory groups in the AIDS program. The general idea was to examine the NIH approach to AIDS research after five years of intensive activity. Headed by William Danforth, chancellor of the University of Washington in St. Louis, the 15-person IOM committee met five times between October 1989 and September 1990 before issuing *The AIDS Research Program of the National Institutes of Health* in 1991.

The report recommended that the period of rapid buildup in AIDS research, which led to expenditures of $805 million in fiscal year 1991, or nearly 10 percent of the total NIH budget, should give way to long-term planning. In terms of emphasis, the committee suggested that NIH should increase its activities in behavioral science, basic science, patient care research, and vaccine development. As William Danforth explained to a congressional committee, major progress depended on "better understanding. The best way to increase this understanding is to provide support for scientists studying how viruses work. . . . Such basic research may not be glamorous but it is our only real hope." In the meantime, according to Danforth, the nation was forced to depend on "behavioral change" for improvements in mortality and morbidity related to AIDS; therefore the country needed "to know more about human behavior and how it might be modified." The committee estimated that implementing its recommendations would lead to an increase of 25 percent in the NIH budget for AIDS but cautioned against siphoning resources from other parts of the research program. Hence, any budget increases should be "new funds."[50]

The 1991 report amounted to a synthesis of the IOM's AIDS concerns, as developed in its 1986 and 1988 reports, and the IOM's traditional interest in improving the management of the NIH and insulating it from politics. The IOM approved of the way in which NIH organized its AIDS activities. Despite the size of the problem, NIH chose not to create an AIDS institute. Instead, NIH managed the program as an "institute without walls" from within the Office of the Director, with an associate director for AIDS research, a national advisory council, an executive committee of senior program representatives, and an executive office for staff support. Such an approach meshed with the recommendations of the 1984 IOM report on the management and organization of the National Institutes of Health, which had cautioned against creating new institutes. In the 1984 report and elsewhere, the IOM had tried to cushion NIH from the yearly vicissitudes of the budget process and from political directives that disrupted the orderly conduct of research. In this spirit, the 1991 report advised that instead of basing the AIDS program on the annual budget process, NIH should develop a five-year plan to set priorities and identify research fields that needed to be strengthened. The IOM study group also suggested, as had the 1984 group, that the NIH director be given more discretion over resource allocation in order to be able to respond to health crises. The committee recommended an annual discretionary fund of $20 million and the authority to transfer up to 1 percent of each NIH appropriation account.[51]

In addition to the study of NIH, the AIDS Oversight Committee took particular interest in the Roundtable for the Development of Drugs and Vaccines Against AIDS. Establishment of the Roundtable, which met for the first time in February 1989, reflected the use of the Institute's convening function. The idea was to "help resolve impediments to AIDS drug and vaccine development" by bringing together representatives of government, academia, the pharmaceutical industry, and patient advocates for regular conversations on pertinent topics. Harold Ginsberg, a professor of medicine and microbiology at Columbia, and Sheldon Wolff cochaired the sessions. Robin Weiss supervised the staff, which was headed by Richard Berzon.[52]

At its first meeting, the group discussed the state of vaccine development, with one conferee noting that the search for a vaccine was "in abeyance," and identified more promising topics for workshops or conferences. The group selected clinical trials as a subject worthy of special discussion, wondering, in particular, about using "prolonged survival" as the sole criterion on which to base FDA approval of new HIV drugs. Perhaps more important "surrogate endpoints," such as particular levels of "p24 core antigen" or particular points on quality of life scales, could be identified. The discussion led to a conference on this subject in September 1989 attended by FDA Commissioner Frank Young and NIH Associate Director for AIDS Anthony Fauci.[53]

On June 26, 1989, the Roundtable used its regularly scheduled meeting to discuss the potential value of consortia in the AIDS drug and vaccine development processes. The discussion took the form of a workshop in response to a congressionally mandated request for information on this topic and led to a formal Roundtable report to the Public Health Service. The group, which consisted of the usual Roundtable members plus four outside experts, decided that the development of animal models, in particular, could benefit from collaborative efforts through research consortia. Such an approach might ease the problem of what one Roundtable member described as an "acute shortage" of animals "to test potential preventive and therapeutic agents against HIV." As for other aspects of the problem, the group was less sure about the value of consortia. Basic biomedical research, for example, had always been a matter for individual researchers and laboratories, and the group could not agree on whether it might be improved through a research consortium. The group could agree, however, on the need for a permanent consortium to examine present and future drug and vaccine issues, and some thought that the National Commission on AIDS might play such a role. As in all the activities of the Roundtable, the conversation was both candid and informal and gave attendees from one sector of the research establishment an opportunity to understand the differing perspectives of those from other sectors.[54]

A few months after the Roundtable meeting on consortia, the IOM held a special event that reflected the interests of both the IOM and the oversight committee in the international dimensions of AIDS. Arranged by the Academy of Medical Sciences of the USSR and the IOM, it consisted of an exchange of information between Soviet and American experts. Thier described it as a "serious working session" in which "we learn about our joint level of knowledge and about the progress of dealing with AIDS in both our countries." Dr. Vadim Pokrovsky of the USSR Ministry of Health told the audience in the NAS auditorium that the first documented case of AIDS in the Soviet Union occurred in 1987. This meant that the Soviet Union was a relative latecomer to AIDS and made it all the more important for Soviet public health authorities to study the American experience. The meeting gave the Soviets a panoramic view of this experience, with talks by the leading American authorities on such subjects as AIDS epidemiology, the treatment of HIV infection, and the prevention of HIV infection. Thier closed the meeting by reminding the participants that the purpose of the U.S.–USSR program was to put American and Soviet scientists in direct contact with one another and expressing confidence that the meeting would produce further interactions and collaborative research efforts. A participant from the U.S. Public Health Service described the results in the optimistic terms of the late Cold War era. If the two "most powerful countries in the world, which so long have been antagonists, can get together to work in areas of benefit to all," it would serve as a "role model for the world," break down "global indifference to the spread of AIDS," and "facilitate worldwide cooperation" in AIDS.[55]

It was, of course, only a meeting, just one of many that the IOM held on AIDS in this period. By this time, the AIDS program had moved well beyond the *Confronting AIDS* projects of 1986 and 1988 to embrace nearly all of the IOM's divisions. In 1989, Robin Weiss and her special AIDS activities branch of the IOM supervised the evaluation of the National Institutes of Health research program, planned a conference to examine the issues surrounding prenatal and newborn screening for HIV infection, and staffed both the AIDS Activities Oversight Committee and the Roundtable for the Development of Drugs and Vaccines Against AIDS. At the same time, the Medical Follow-Up Agency was planning to locate and study a group of HIV-positive servicemen who had been lost to follow-up after discharge. The Division of International Health hoped to establish an International Forum for AIDS Research, and the Division of Health Promotion and Disease Prevention had formed a subcommittee to advise the American National Red Cross on AIDS program materials.[56]

If the job of the AIDS Oversight Committee was to stimulate activity, then the committee could count its work a success. In June 1991, the IOM disbanded its special unit on AIDS activities and moved its remaining projects to the Division of Health Promotion and Disease Prevention, noting that the move "did not indicate a diminished interest in AIDS-related projects at the IOM, which will be pursued as vigorously as in the past." By this time, thanks to the work of the AIDS Oversight Committee and its predecessor, an interest in AIDS was woven into the basic fabric of the IOM.[57]

Continuing Interest in AIDS

Two projects in particular demonstrated the continuing interest of the IOM in AIDS. One involved a large-scale study of the transmission of HIV through the blood supply in the early 1980s, and the other surveyed the AIDS research programs of what was known before 1992 as the Alcohol, Drug Abuse, and Mental Health Administration. Both studies stemmed from congressional mandates and reflected the traditional IOM mission to provide oversight and guidance to federal agencies. The fact that Congress asked the IOM to do these studies illustrated its confidence in the Institute's impartial judgment and its respect for the IOM's expertise on AIDS.

The project on HIV and the blood supply resulted from one of the many tragedies of the AIDS epidemic. In the early 1980s, as a result of using blood-based products, half of the 16,000 hemophiliacs in the United States contracted the AIDS virus. Before the development of an effective blood test for HIV, the introduction of HIV screening kits, and the routine use of antiviral heat treatment for blood products in

1985, more than 12,000 recipients of blood transfusions also became infected with HIV. This situation raised a host of questions and led to a search to assign blame for this apparent breakdown in the nation's system of public health protections. Hemophiliacs, in particular, lobbied Congress in an effort to affix blame and seek redress.[58]

In April 1993, Senators Edward Kennedy (D-Mass.) and Bob Graham (D-Fla.) and Congressman Porter Goss (R-Fla.) requested that the Department of Health and Human Services investigate what had happened. Responding in July 1993, Secretary of Health and Human Services Donna Shalala agreed on the need for an investigation. For her and the Democratic party, such an investigation offered an opportunity to delve into an apparent mistake that had been made by a Republican administration. On the surface, at least, it looked as though this mistake stemmed in part from an overzealous reliance on the private market to maintain the public's safety. Because Republicans had allowed the CDC and the FDA to deteriorate as a result of a predetermined belief in the futility and inefficiency of federal regulation, a tragedy of epidemic proportions had occurred. Furthermore, the Democrats in Congress and in the executive branch realized that hemophiliacs and recipients of blood transfusions were, along with children who had been infected by their mothers, the perfect victims of this tragedy to exploit for political purposes. Unlike homosexuals who engaged in anal intercourse or drug abusers who injected poison into their veins, neither hemophiliacs nor blood transfusion recipients, such as tennis player Arthur Ashe, could be said to have brought AIDS on themselves. Aware of these political overtones, Shalala made sure to cast the project not as a retrospective witch hunt but rather as a future-oriented effort to draw lessons that might be relevant to future threats to the safety of the nation's blood supply. To make sure it was not labeled the product of a partisan administration, she commissioned the Institute of Medicine to do the study, which got under way in 1994.[59]

Harold C. Sox, Jr., head of the Department of Medicine at Dartmouth, chaired the 14-person committee. Half of the people on the committee were physicians. The other half consisted of a nurse, an ethicist, one of the nation's leading historians of medicine, a prominent student of administrative law, a political scientist who had written widely on public administration and bureaucratic behavior, a sociologist, and a lawyer. It was an extremely able and intellectually versatile group, with no concessions made to the many interest groups who had pushed Congress to investigate the matter. Not until April 1994 did this group begin to meet in a serious way. The composition of the committee and the delay in initiating the project led to a certain amount of restiveness in Congress. Although Congressman Goss and Senator Graham called attention to the NAS's "integrity and professionalism in conducting complex and sensitive studies," they shared the fears of their constituents that the study "would drag on too long and yield inconclusive results."[60]

To do its job, the committee had to delve into the decisionmaking processes that governed the control of the nation's blood supply between 1982 and 1986. This required the committee to assemble an archive of documents from government agencies such as the Food and Drug Administration, consumer interest groups such as the National Hemophilia Foundation, private groups involved in the blood supply system such as the American Association of Blood Banks, and suppliers of blood and blood products such as the American Red Cross. To make sense of these documents, the committee studied the nation's complex system of acquiring, distributing, and regulating the quality of its blood supply. The committee had to determine how consumers of blood products, the doctors who cared for these consumers, and public health authorities came to be aware of AIDS and of the fact that HIV could be transmitted through blood. The events between December 1982, when the first report of AIDS in a patient with hemophilia appeared, and the 1985 decision to undertake routine screening of blood for HIV were particularly crucial. As the committee tried to understand how the system worked and what had caused it to go wrong, it received a great deal of help from study director Lauren Leveton and from others who worked for the IOM's Division of Health Promotion and Disease Prevention.[61]

From the beginning, the committee heard from the victims of tragedy. On May 16, 1994, it met with Dana Kuhn, who spoke on behalf of the Committee of Ten Thousand, a national organization that provided peer advocacy and support to persons infected with HIV through tainted blood and blood products. On March 26, 1983, Kuhn, a medical doctor and hemophiliac, received HIV- and hepatitis-contaminated factor VIII, one of the naturally occurring proteins in the liquid part of the blood (known as plasma) that aided coagulation. He developed AIDS and, before he became aware of his condition, transmitted HIV to his wife through heterosexual intercourse. She died in 1987, leaving him with a 6-year-old and a 4-year-old to raise and with the expectation that he too would soon die. The committee also heard from Mr. Richard Valdez, the president of the Peer Group Association, who spoke about the death of one of his sons and the terminal illness of the other, as a result of using a blood product known generically as a "factor concentrate." In September, the committee heard the testimony of Jonathan Wadleigh. "I have severe hemophilia and AIDS," he said; "I lost my brother to hemophilia-associated AIDS in 1986." He proceeded to relate his story, telling the committee that he had never had life-threatening bleeding due to hemophilia but that he had decided to use antihemophilic products to support his "active lifestyle." According to Wadleigh, "I was never told about the risks associated with the use of antihemophilic factor concentrates or given a choice as to what products I used. My experience is typical."[62]

As the evidence mounted, the committee had to decide what to make of it. The dilemmas that the committee faced were those of the historian and the policy analyst. The evidence was, by its very nature, ambiguous—the product of faulty memories and differing perceptions of reality. Although contemporary documents helped to firm up times and dates, they too were subject to interpretation. Bureaucratic memoranda, for example, were written for many purposes, for example, to put forward a point of view or shape the course of future events, not merely to set down events for the record. Reading the documents, the committee had constantly to keep in mind that the participants in the events of the early 1980s did not know how things would turn out. People acted on the basis of available knowledge, which in retrospect could be shown to be flawed. Once the committee agreed on a version of what happened, it still had to tease out the significance. The danger was that the committee would base its recommendations for future actions on the basis of past behavior, without allowing policymakers to explore the similarities and differences between a past event and a future occurrence. In other words, the committee had to guard against recommending that the public health community fight the last war. The problems that the committee faced showed just how hard it was to create a usable past and, in general, how difficult it was to come to terms with the AIDS epidemic.[63]

In the July 1995 report, the committee presented a nuanced, balanced history of decisionmaking during the AIDS crisis and made reasoned recommendations designed to improve the nation's performance in a future crisis. The committee concluded that the events it analyzed "underscore the difficulty of personal and institutional decisionmaking when the stakes are high, when knowledge is imprecise and incomplete, and when decisionmakers may have personal or institutional biases." Whatever the difficulties, it was clear in retrospect that the system "did not deal well" with blood safety issues. The committee decided that "unless someone from the top exerts strong leadership, legal and competitive concerns may inhibit effective action" by federal agencies. Further, the Food and Drug Administration lacked "a systematic approach to conducting advisory committee processes" and relied too heavily on the entities it regulated for data analysis. The lack of both strong leadership and an effective advisory committee process inhibited public agencies from thinking ahead and led to a cautious, and ultimately destructive, response to the crisis.

To remedy these perceived problems, the committee made 14 specific recommendations. Some were bureaucratic, such as the suggestion that the Secretary of Health and Human Services "designate a Blood Safety Director at the level of deputy assistant secretary or higher" or that the Public Health Service "establish a Blood Safety Council." Others were strategic, such as the recommendation that federal agencies support and respond to the CDC in its responsibility to "serve as the nation's early warning system for threats to the health of the public" or that the FDA encourage the blood industry "to implement partial solutions that have little risk of causing harm" in cases where "uncertainties or countervailing public health concerns preclude completely eliminating potential risks." The committee also cautioned physicians and patients "faced with a decision in which all options carry risk" to "take extra care to discuss a wide range of options." To facilitate the flow of reliable information, the committee recommended that the Department of Health and Human Services convene an expert panel to inform people about the "risks associated with blood and blood products" and about the safety and efficacy of available alternatives.[64]

The Committee on HIV and the Blood Supply did exactly what Secretary Shalala hoped it would. It stepped away from the emotional context of the situation and produced a dispassionate, yet thorough, analysis of the problem. It also made recommendations that showed sophistication about the ways in which bureaucracies and the political process worked. Secretary Shalala responded to these recommendations almost immediately, designating the Assistant Secretary for Health as the DHHS Blood Safety Director and creating a Blood Safety Committee composed of the heads of the FDA, CDC, and NIH. The report revealed the effectiveness of letting people with differing intellectual perspectives examine a problem in order to arrive at a useful synthesis. The ability to assemble a group with a wide range of interests and capabilities to examine a controversial problem without fear of political reprisal constituted a great strength of the IOM. The report on HIV and the blood supply not only showed these attributes to best advantage but also demonstrated that the

IOM and its staff had developed a sort of collective expertise on the subject of AIDS that could be put to a variety of uses.[65]

In 1994, the Institute of Medicine published a report on substance abuse and mental health issues in AIDS research that resulted from a congressional request for the IOM to investigate the AIDS-related research programs of the Alcohol, Drug Abuse, and Mental Health Administration (ADAMHA). In making this request, Congress specified that the evaluation be similar to the one the IOM had already performed for the National Institutes of Health. The final report contained references to no fewer than six previous IOM reports that were relevant to the specific subject of AIDS and behavior. H. Keith H. Brodie, president emeritus and James B. Duke Professor of Psychiatry at Duke University, chaired the advisory committee. A reorganization in the Department of Health and Human Services that took effect on October 1, 1992, made his job particularly difficult. Under this reorganization, the National Institute on Alcohol Abuse and Alcoholism, the National Institute on Drug Abuse, and the National Institute of Mental Health moved to the National Institutes of Health; the service-providing components of ADAMHA became part of the Substance Abuse and Mental Health Services Administration. In addition, a bill passed in July 1993 changed the allocation of money for AIDS research within the National Institutes of Health. Hence, the committee faced a moving target.

As with the study of HIV and the blood supply, the Committee on Substance Abuse and Mental Health Issues in AIDS research found a slow response on the part of the federal bureaucracy to the AIDS crisis. In the meantime, AIDS continued to grow as a public health problem. By the end of 1992, AIDS was the leading cause of death among men 25–44 years of age. Not until 1987 did the National Institute on Alcohol Abuse and Alcoholism devote more than 1 percent of its expenditures to AIDS research, and similar patterns applied to the other NIH institutes under study. When these institutes did start to spend a substantial amount of money on AIDS research in 1987, such as $139.3 million by the National Institute on Drug Abuse, they concentrated most of it on biological research and neglected behavioral research. Part of the problem was a political taboo against such activities as a "federally sponsored, national survey of sexual behavior to help determine the nature and level of risk for HIV transmission in the general population." The committee found numerous other gaps, for example, describing the effect of AIDS on those already suffering brain disorders as "remarkably understudied," and offered specific suggestions on how to close the gaps. It recommended that the three NIH institutes study how individuals from diverse backgrounds "cope with the reality of having family members who are infected with HIV." It also suggested that they develop new programs "to encourage and facilitate innovative, collaborative, and cross-disciplinary proposals." To make sure AIDS received the proper emphasis at these institutes, the committee recommended that each establish "a full-time AIDS coordinator." Each institute should also "develop initiatives to support research on the role of social, cultural, and structural factors in HIV/AIDS transmission, prevention, and intervention."[66]

Of necessity, *AIDS and Behavior* was a diffuse document, in part because of the original congressional mandate and in part because the subject did not lend itself to an overarching theme. Unlike the 1986 report and its 1988 update, the behavior study did not represent a general call to arms as much as a fine-tuning of bureaucratic priorities. The presence of such a study showed how the IOM addressed two substantively different audiences on the subject of AIDS. On the one hand, it sought to address the general public and the top levels of political leadership through studies such as *Confronting AIDS*. On the other hand, it attempted to reach the professional research community and the inner levels of the federal bureaucracy through studies such as *AIDS and Behavior*. Both groups listened.

The Institute and AIDS

In the decade between 1985 and 1995, AIDS became one of the IOM's signature activities, as Samuel Thier had hoped it would. The Institute could take credit for urging federal action to combat the epidemic at a time when the top levels of government preferred to ignore it. The 1986 and 1988 reports set the stage for the nation's response to AIDS and made the Institute of Medicine visible to the public for the first time. Many of the recommendations in these reports were written into law by Congress or put into operation by the federal bureaucracy. In August 1990, for example, Congress passed the Ryan White Comprehensive AIDS Resources Emergency Act, which established a federal grant program to cities that were affected by the epidemic. Congress patterned the program after the one that Stuart Altman had devised and the Oversight Committee on AIDS had recommended in 1988. The very existence of the congressionally mandated National Commission on AIDS, which served from 1989 to 1993, could be traced to recommendations in the 1986 report. In 1991, the Office of AIDS Research in the National Institutes of Health issued a plan for AIDS research that marked a response to the IOM's report on AIDS research at the NIH. On the subject of AIDS, therefore, the IOM's recommendations brought strong responses.

During the discussion of the Sproull report, a foundation executive had remarked that if the IOM did not exist, something similar would have to be invented to take its place. Whether or not this was true at the time remains a matter of debate. In the case of AIDS, however, the statement carried an undeniable element of truth. Simply put, without the IOM, the federal government's response would certainly have been delayed and, in all likelihood, different. The Institute of Medicine succeeded in letting the nation hear the opinions of doctors and scientists in the CDC, NIH, and elsewhere whose advice was largely ignored at the upper levels of the bureaucracy. Through the offices of the Institute of Medicine, these physicians and scientists joined their academic and private-sector counterparts in demanding that more attention be paid to the science, epidemiology, and patient care aspects of AIDS. In the case of AIDS, the IOM made a difference.

The IOM could not have played such a prominent role on AIDS if Samuel Thier had not won the respect of Frank Press. Because Press trusted Thier, he allowed the IOM to take the lead on an important aspect of science policy. Thier earned this respect in part through efficient management, in part through his ability to raise money, and in part because of his understanding of how scientists approached a problem. As a result, the AIDS activities of the IOM were true collaborations between the IOM and the NAS. The fact that Nobel Prize-winning scientists and the nation's leading medical authorities both contributed to the 1986 and 1988 reports made these reports all the more effective.

The AIDS activities of the IOM created their own sense of momentum. General, overarching reports led to requests that the IOM organize workshops, create forums, and conduct studies on more specialized and focused topics. This meant that the IOM's reputation in the field of AIDS was not tied to Samuel Thier and could continue under his successors. This was fortunate, because the size of the epidemic made it necessary for the IOM to have a continued involvement in activities related to AIDS. As the IOM's Committee on Substance Abuse and Mental Health Issues pointed out in 1994, the rate of increase in AIDS cases was alarming. The first 100,000 reported cases occurred within an eight-year period; the second, in a two-year period. As the IOM entered its second quarter century, the AIDS epidemic remained a constant presence in American life.

Notes

1. Quoted in Eve K. Nichols, *Mobilizing Against AIDS: The Unfinished Story of a Virus* (Cambridge, Mass.: Harvard University Press, 1986), p. 3.

2. Allen M. Brandt, "AIDS: From Social History to Social Policy," in Elizabeth Fee and Daniel M. Fox, eds., *AIDS: The Burdens of History* (Berkeley: University of California Press, 1988), p. 153.

3. IOM Council Meeting, Minutes, July 20, 1983, Institute of Medicine (IOM) Records, National Academy of Sciences (NAS) Archives; Allen M. Brandt, "AIDS: From Social History to Social Policy," p. 161.

4. IOM Council Meeting, Minutes, March 18, 1985, IOM Records.

5. IOM Council Meeting, Minutes, September 23, 1985, IOM Records; Centers for Disease Control, *Morbidity and Mortality Weekly Report* 34 (August 30, 1985), pp. 517–521.

6. Institute of Medicine, *Annual Report 1985, Program Plan 1986* (Washington, D.C.: National Academy Press), p. 4; Eve K. Nichols, *Mobilizing Against AIDS*, pp. 197–198.

7. Eve K. Nichols, *Mobilizing Against AIDS*, pp. 2–3, 59.

8. IOM Council Meeting, Minutes, November 18, 1985, IOM Records.

9. *Ibid.*

10. See front and back matter in Institute of Medicine, *Confronting AIDS: Directions for Public Health, Health Care, and Research* (Washington, D.C.: National Academy Press, 1986).

11. "Develop a National Agenda on Research and Health Care for AIDS," January 22, 1986, IOM Development Files, Accession 91-045, IOM Records.

12. Transcript of Meeting of the Epidemiology Workshop, March 6, 1986, pp. 4, 10, 12, 14, Accession 95-065, IOM Records.

13. Transcript of Plenary Session, March 13, 1986, pp. 60, 106, 126, Accession 95-065, IOM Records.

14. Transcript of the Public Health Panel Meeting, October 13, 1986, pp. 87–88, Accession 95-065, IOM Records.

15. Proceedings of the Research Panel, March 13, 1986, pp. 42, 112, 117, 121, 126, 129, 150, IOM Records.

16. Transcript of Public Health Panel Meeting, March 13, 1986, pp. 2, 11, 14–19, 35, 39, 75, 82, Accession 95-065, IOM Records.

17. Transcript of Plenary Session, March 14, 1986, pp. 198, 242, 248, 265, Accession 95-065, IOM Records.

18. Agenda for Meeting on Research, Health Care, and Public Health Strategies, New York Blood Center, May 15, 1986, and Agenda of Meeting of Committee on a National Strategy for AIDS, April 8–9, 1986, both in Accession 91-065, IOM Records.

19. "AIDS Research Commission Proposal—Draft," June 2, 1986, Accession 95-065, IOM Records.

20. "Blue Ribbon Panel Urges Greatly Expanded Education and Research Effort Against AIDS," IOM Press Release, October 29, 1986, Accession 91-045, IOM Records; "Confronting AIDS: Directions for Public Health, Health

Care, and Research, Summary of Recommendations," Accession 95-070, IOM Records.

21. *Confronting AIDS*, pp. 5–37.

22. *Ibid.*, p. 19.

23. *Ibid.*, pp. 9–12.

24. Philip M. Boffey, "Federal Efforts on AIDS Criticized as Gravely Weak," *New York Times,* October 30, 1986, p. 1.

25. Christine Russell, "Escalate AIDS Fight, Scientists Urge," *Washington Post,* October 30, 1986, p. A-1.

26. Network Television Evening News Abstracts for October 29, 1986, Vanderbilt Television News Archive, Vanderbilt University.

27. Frank Press and Samuel Thier to David Baltimore and Sheldon Wolff, December 17, 1986, Accession 95-065, IOM Records.

28. IOM Council Meeting, Minutes, November 17–18, 1986, and July 23–24, 1987, IOM Records.

29. Paul Rogers to Samuel Thier, September 3, 1987, Accession 95-066, IOM Records.

30. Institute of Medicine, *An Agenda for AIDS Drug Development: Report on the Conference on Promoting Drug Development Against AIDS and HIV Infection* (Washington, D.C.: National Academy Press, 1987), pp. 1, 24; Transcript of presentation by David Baltimore, August 31, 1987, Accession 95-066, IOM Records.

31. "Evaluation of the Modeling of the Spread of Infection with Human Immunodeficiency Virus and the Demographic Impact of Acquired Immune Deficiency Syndrome Worldwide," August 20, 1987, Accession 95-064, IOM Records; Roy Widdus to Frank Press, November 30, 1987, Accession 95-064, IOM Records; Institute of Medicine, *Approaches to Modeling Disease Spread and Impact: Report of a Workshop* (Washington, D.C.: National Academy Press, 1988), p. 1, 5–7; *Confronting AIDS,* p. 31.

32. Samuel Thier to Gerald Mossinghoff, President, Pharmaceutical Manufacturers Association; Peter J. Fischinger, Deputy Director, National Cancer Institute, to Thier, August 8, 1987; Roy Widdus to Frank Press, November 19, 1987; and Transcript of David Baltimore remarks, December 14, 1987, all in Accession 95-068, IOM Records.

33. Institute of Medicine, *Prospects for Vaccines Against HIV Infection: Report of the Conference on Promoting Development of Vaccines Against Human Immunodeficiency Virus Infection and Acquired Immune Deficiency Syndrome* (Washington, D.C.: National Academy Press, 1988), pp. 15, 20, 23.

34. Roy Widdus to Frank Press, November 30, 1987, AIDS Oversight Committee Files, Accession 95-070, IOM Records.

35. Robin Weiss, "Notes on AIDS Oversight Committee Meeting," October 24, 1987, AIDS Oversight Committee Files, Accession 95-070, IOM Records.

36. IOM Council Meeting, Minutes, November 16–17, 1986, IOM Records; Samuel Thier to Robert E. Windom, September 17, 1987; Thier to Leighton Cluff, October 16, 1987; and Cluff to Thier, November 5, 1987, all in Accession 95-050, IOM Records.

37. Robin Weiss, "Notes on AIDS Oversight Committee Meeting."

38. "Recommendations from December 21, 1987, Meeting with Brief Discussion," Accession 95-070, IOM Records.

39. Robin Weiss, "Notes on AIDS Oversight Committee Meeting."

40. "Recommendations from December 21, 1987, Meeting."

41. *Ibid.;* Stuart Altman to Robin Weiss, January 12, 1988, Accession 95-050, IOM Records.

42. "Panel Cites Remaining Deficiencies in National Effort to Combat AIDS," IOM Press Release, June 1, 1988, Accession 95-070, IOM Records; Institute of Medicine, *Confronting AIDS: Update 1988* (Washington, D.C.: National Academy Press, 1988).

43. Philip M. Boffey, "Expert Panel Sees Poor Leadership in U.S. AIDS Battle," *New York Times,* June 2, 1988, p. A-1; William Booth, "AIDS Panel Converges on a Consensus," *Science,* June 10, 1988, pp. 1395–1396.

44. Philip M. Boffey, "Expert Panel Sees Poor Leadership in U.S. AIDS Battle"; "Making a Difference," July 24, 1989, Yordy Files, Accession 91-051, IOM Records.

45. "HIV Infections and AIDS: Recommendations to the President-Elect," Accession 95-070, IOM Records.

46. *Ibid.*, p. 2.

47. Frank Press and Samuel Thier to Honorable George Bush, December 13, 1988, Accession 95-070, IOM Records.

48. Monroe Trout to Mrs. George Bush, December 14, 1988; Trout to Samuel Thier, December 16, 1988; Thier to Trout, January 3, 1989; and George Bush to Thier and Frank Press, January 10, 1989, all in Accession 95-070, IOM Records.

49. Notes, AIDS Oversight Committee Meeting, October 3, 1988, Accession 95-050, IOM Records.

50. Anthony Fauci, Associate Director for AIDS Research, National Institutes of Health, to Robin Weiss, August 22, 1988, Accession 95-050, IOM Records; Weiss to Representative Henry A. Waxman, April 11, 1991, Accession 95-070, IOM Records; Testimony of William Danforth before the Human Resources and Intergovernmental Relations Subcommittee of the Committee on Government Operations, March 7, 1991, Accession 95-070, IOM Records; Institute of Medicine, *The AIDS Research Program of the National Institutes of Health* (Washington, D.C.: National Academy Press, 1991).

51. Testimony of Melvin M. Grumbach, Edward B. Shaw Professor of Pediatrics, University of California at San Francisco, before the Human Relations and Intergovernmental Relations Subcommittee, March 7, 1991, Accession 95-070, IOM Records.

52. Robin Weiss to Phil Smith, November 17, 1989, Accession 95-070, IOM Records; "Roundtable for the Development of Drugs and Vaccines Against AIDS, Meeting Summary," February 17, 1989, Accession 95-050, IOM Records.

53. *Ibid;* "Surrogate Endpoints in Evaluating the Effectiveness of Drugs Against HIV and AIDs," September 11–12, 1989, Accession 89-013-5, IOM Records.

54. "Roundtable for the Development of Drugs and Vaccines Against AIDS, Meeting Summary." June 26, 1989, Accession 95-050, IOM Records; "The Potential Value of Research Consortia in the Development of Drugs and Vaccines Against HIV Infection and AIDS: Report of a Workshop," 1989, Accession 95-070, IOM Records.

55. "Summary of the Institute of Medicine U.S.–USSR AIDS Symposium, October 4–5, 1989," February 28, 1990, Accession 95-069, IOM Records; Transcript of Samuel Thier's remarks, October 4, 1989, Accession 95-069, IOM Records; Peter Hartwick to Robin Weiss, November 6, 1989, Accession 95-069, IOM Records.

56. *Institute of Medicine Annual Report 1989,* pp. 59–60.

57. *Institute of Medicine Annual Report 1991,* p. 35.

58. "Proposed Study Plan," June 15, 1994, Records of the Committee to Study HIV Transmission Through Blood or Blood Products, and "Working Group Meeting Summary," April 5, 1994, both in Accession 96-004, IOM Records.

59. Senator Bob Graham and Representative Porter Goss to Michael Stoto, January 24, 1994, Accession 96-004, IOM Records.

60. Representative Porter Goss and Senator Bob Graham to Representative John Dingell, January 18, 1994, and "Committee Appointment," September 30, 1993, both in Accession 96-004, IOM Records.

61. Lauren Leveton, Ph.D., to Reid Stuntz, Staff Director and Chief Counsel, House Subcommittee on Oversight and Investigations, Committee on Energy and Commerce, December 9, 1994, and "Working Group Meeting Summary," April 5, 1994, both in Accession 96-004, IOM Records.

62. "Minutes," May 16, 1984, and Transcript of Proceedings of Public Hearing, September 12, 1994, both in Accession 96-004, IOM Records.

63. Daniel Fox and Elizabeth Fee, eds., *AIDS: The Burdens of History* (Berkeley: University of California Press, 1988); Richard Neustadt and Harvey Fineberg, *The Swine Flu Affair: Decisionmaking on a Slippery Disease* (Washington, D.C.: U.S. Department of Health, Education, and Welfare, 1978).

64. Institute of Medicine, Lauren B. Leveton, Harold C. Sox, Jr., and Michael A. Stoto, eds., *HIV and the Blood Supply: An Analysis of Crisis Decisionmaking* (Washington, D.C.: National Academy Press, 1995), pp. 5–17.

65. "Testimony of Donna E. Shalala at House Committee on Government Reform and Oversight, Subcommittee on Human Resources and Intergovernmental Relations," October 12, 1995, Accession 96-004, IOM Records.

66. Institute of Medicine, Judith D. Auerbach, Christina Wypijewska, and H. Keith H. Brodie, eds., *AIDS and Behavior: An Integrated Approach* (Washington, D.C.: National Academy Press, 1994), pp. v, vi, 1, 2, 15, 19, 23, 26–29, 31–33.

7

The Institute of Medicine at Twenty Five

When Philip Handler declared the Institute of Medicine (IOM) open for business in 1970, he did not receive a communication from the President of the United States. Although the start of the IOM occasioned comment in journals such as *Science* and among those who followed the affairs of the National Academy of Sciences (NAS), it failed to elicit much reaction from the external policy community. Roger Bulger, the IOM's first executive officer, stated in 1974 that the IOM had not become a household word or, he might have added, a major player in health care policy. Some 21 years later, this situation had clearly changed. When the Institute of Medicine celebrated its first 25 years in December 1995, it heard from Health and Human Services (HHS) Secretary Donna Shalala, who thanked the Institute for "giving her so many health care leaders." She referred to the fact that nearly all of the agencies in the department concerned with health were headed by IOM members such as David Kessler at the Food and Drug Administration (FDA) and Bruce Vladeck at the Health Care Financing Administration (HCFA). Shalala also brought along a special message from President Bill Clinton. "Since 1970," the President wrote, "the Institute of Medicine has contributed thoughtful and wise health policy covering an extraordinary range of issues."[1]

Kenneth Shine Arrives

It was Kenneth Shine who presided over the anniversary celebration. The dean of the University of California at Los Angeles (UCLA) Medical School, he interviewed for the IOM job during the 1991 annual meeting. The Council wanted someone to take over from Samuel Thier, who had announced his departure in May. As the search proceeded during the fall of 1991, Stuart Bondurant, dean of Medicine at the University of North Carolina at Chapel Hill, filled in on an interim basis. At the end of the year, the Institute of Medicine announced the formal appointment of Shine, who assumed the job on a full-time basis on July 1, 1992.

In selecting Shine, the Institute of Medicine preserved the tradition of having a leading academic health administrator as its president. Although Shine grew up in modest circumstances in Providence, Rhode Island, a special scholarship for talented students of the Providence public schools enabled him to go to Harvard College and then to Harvard Medical School. Like Samuel Thier, Kenneth Shine took his internship and residency at the Massachusetts General Hospital, another institution with a Harvard affiliation. At Massachusetts General, he developed an expertise in the muscles of the heart and sought an appropriate postdoc laboratory experience. He found it at UCLA and remained there as he worked his way up the academic and administrative ranks to become dean. Along the way, he developed expertise in the fields of cardiology, physiology, and academic administration, and as a result of his work with the Association of American Medical Colleges and the American Heart Association, he also became interested in public policy.

Kenneth Shine inherited a large and complex organization from Samuel Thier. Each division produced numerous reports every year, and each division also played a part in the IOM's convening function by administering a special committee, forum, lecture series, or award. Enriqueta Bond, the IOM's executive officer, did a masterful job of keeping these activities on track during the year between Thier's departure and Shine's arrival on a full-time basis. When Shine took over the daily operations of the IOM in the summer of 1992, the country was in the middle of an election campaign that would bring the Democrats back into power. One of the key decisions that Shine faced was how to position the Institute of Medicine in an era of Democratic rule.

The Inner Details of Health Care Policy

As Shine contemplated how to approach his presidency, each division continued the work that led to reports on a wide variety of subjects. For example, a project on the implications of regional health care data bases for health policy fell into the bailiwick of the Division of Health Care Services, still headed by Karl Yordy, who had come to the IOM in the era of John Hogness and would remain until his retirement in October 1993. Roger Bulger, another old IOM hand whose arrival at the IOM had preceded Yordy's, chaired the steering committee that produced a report on *Health Data in the Information Age* in 1994. Similar in tone to an earlier report on the computer-based patient record, the study tried to establish some guidelines for the use of health data bases that respected a patient's right to privacy yet yielded appropriate and helpful data. The committee realized that the information contained in the data bases could serve two conflicting purposes. On the one hand, the information might violate an individual's privacy and produce adverse consequences; on the other hand, however, it had the potential to improve the quality of medical care for large groups of individuals. To minimize the harm and maximize the benefits, the steering committee recommended that employers "not be permitted to require receipt of an individual's data from a health database organization as a condition of employment or the receipt of benefits" but suggested that "health database organizations should release non-person-identifiable data upon request to other entities once those data are in analyzable form."[2]

Another project from the Division of Health Care Services came from a congressional mandate that the Institute of Medicine study whether there was a need for more nurses in hospitals and nursing homes in order to improve the quality of patient care and reduce work-related injuries and stress among nurses. Nursing professional associations and special interest groups, concerned about changes in health care settings made to promote efficiency and reduce costs, supplied the impetus for the study. Receiving a formal request from the Health Resources and Services Administration to undertake the study in 1994, the Institute of Medicine assembled a committee headed by Frank Sloan, an economist from Vanderbilt, and Carolyne Davis. The choice of Davis, an experienced nurse who had gone into the field of nursing education and ultimately became the head of HCFA, was particularly appropriate. Davis and Sloan, assisted by IOM staff member Gooloo Wunderlich, conducted an investigation that lasted two years and included literature reviews, public hearings, and site visits.

In a report released in January 1996, the committee concluded that the number of registered nurses was adequate to meet national needs. It nonetheless made many specific recommendations for necessary changes in staffing patterns and for research projects needed to refine public policy decisions. In particular, the committee recommended that the National Institute of Nursing Research fund research on "the relationships between quality of care and nurse staffing levels." It suggested that Congress set as a goal for the year 2000 a requirement that there be 24-hour coverage by registered nurses in nursing facilities (the current standard was 8 hours). The committee felt that geriatric nurse specialists should be more involved in the supervision of nursing care and in the direct provision of such care. Hospitals and nursing homes, for their part, should carefully screen applicants for patient care positions filled by nurse assistants to make sure they had no past history of patient abuse.[3]

Both of these studies exemplified the traditional IOM role of appointing an expert panel to help in the process of setting standards of medical practice and administration and to identify topics that required more research. They also helped validate the conjectures of health policy researchers, for example, in the finding of the nursing study that the quality of hospital care had not suffered after implementation of the prospective payment system under Medicare.[4] In this capacity, the studies served the additional function of laying a controversy to rest by virtue of examining the available evidence in an impartial manner. Even in an IOM study, however, there were possibilities for what might be described as latent conflicts of interest. A panel led by Carolyne Davis passed judgment on a prospective payment system that she had helped to put in place as the administrator of the Health Care Financing Administration. Panel members issued calls for more research that they might end up doing themselves.

Neither of the studies had a particularly compelling theme or message. Not to put too fine a point on it, they were not the sorts of publications one might read for pleasure. People were likely to read the reports only if they had a direct interest in the subject at hand. The Institute of Medicine did these sorts of studies either in direct response to a congressional inquiry or because of an interest on the part of a foundation or a component of the medical community. The existence of such studies furthered the IOM's reputation as an organization concerned with what might be described as the inner details of health care policymaking.

The Institute as Public Health Crusader

In the era of Kenneth Shine, the Institute of Medicine pursued these routine studies, yet it lost none of the crusading spirit in the field of public health that could be traced back to Walsh McDermott. The work of the Division of Biobehavioral Sciences and Mental Disorders (renamed the Division of Neuroscience and Behavioral Health in 1996) on *Growing Up Tobacco Free,* a 1994 report, showcased this spirit. Unlike the reports on nursing or regional data bases, this one received an enthusiastic launch from the Institute of Medicine. This meant issuing a separately bound summary and taking extra care with the report's design. A poster, drawn by Juan Morrell, a young student at PS 54 in Brooklyn, appeared on the cover. "I have a dream," the poster read in an obvious reference to Martin Luther King, Jr., and the promises of the civil rights movement, of "a beautiful world without cigarettes." The poster's design featured a neighborhood, a clean and well-lighted place in which people walked the streets safe and secure and free of the menace of cigarette smoke. Although the poster reflected a child's fantasy view of the world, it also reinforced the theme of the report. The report took the harmful nature of cigarette smoking for granted. The problem for public health authorities fell into the domain of behavioral modification: how to stop people from smoking. The basic insight of the report was that antismoking campaigns should be targeted at young people. "In the long run, tobacco use can be most efficiently reduced through a youth-centered policy aimed at preventing children and adolescents from initiating tobacco use," the report stated. Here, then, was a compelling report with a clear message.[5]

The report contained hard-hitting recommendations on ways to restrict the use of tobacco products to adults. The committee, headed by Paul Torrens of UCLA's School of Public Health, proposed banning tobacco products from public vending machines as well as a ban on the free distribution of tobacco products at public events or through the mail. At the heart of the committee's strategy to prevent young people from starting to smoke was a mass media campaign that would include explicit antitobacco advertisements on commercial television stations. To help finance this campaign and further to discourage smoking, the committee suggested an increase in the excise tax on cigarettes. The report accompanied these recommendations with charts and graphs that eloquently summed up the case against smoking. One chart, for example, made the point that "cigarettes kill more Americans than AIDS, alcohol, car accidents, murders, suicides, drugs and fires combined."[6]

In a sense, *Growing Up Tobacco Free,* funded by the Robert Wood Johnson Foundation, constituted a free ride for the Institute of Medicine. No one in the public health community advocated the use of tobacco. To recommend an antismoking campaign reinforced feelings people already had; it was the public health equivalent of a bipartisan good government measure. Although the tobacco companies exerted substantial influence over Congress and state legislatures, they were less of a factor in the communities that supported the IOM and from which the IOM drew its members and audience. By way of contrast, two projects of the Division of Health Promotion and Disease Prevention, one on Agent Orange and the other on unintended pregnancy, took the IOM farther into the realm of controversy.

The Division of Health Promotion and Disease Prevention functioned as one of the IOM's busiest divisions during the presidency of Kenneth Shine. The division administered many of the IOM's convening functions such as the National Forum on the Future of Children and Families, the National Forum on Health Statistics, and the Roundtable for the Development of Drugs and Vaccines Against AIDS. It also assisted in the process of selecting the winner of the Gustav O. Lienhard Award, a prestigious prize that the IOM conferred on accomplished members of the health community for achievement in improving personal health care services. Often this award went to people who had played a prominent role in the development of the IOM itself. In 1993, for example, David E. Rogers, the Walsh McDermott University Professor of Medicine at Cornell University and previously the head of the Robert Wood Johnson Foundation, received the award. In other years, the award went to Robert Ball, the former Social Security commissioner and resident scholar at the Institute of Medicine, and Julius Richmond, President Carter's Surgeon General and "pro tem" head of the Institute of Medicine between Donald Fredrickson and David Hamburg.

Like other Institute of Medicine components, the staff of the Division of Health Promotion and Disease Prevention spent most of its time not on convening or honorific functions but on the conduct of actual studies, such as the one begun in 1992 on Agent Orange. The study stemmed from a congressional request, contained in a law passed at the beginning of 1991, for the National Academy of Sciences to conduct a study of the health effects of Agent Orange and other herbicides used during the Vietnam War. The military used these herbicides as a way of clearing the underbrush that obscured the targets of aerial bombing or provided cover for enemy soldiers intent on killing American and South Vietnamese forces. Although most of the herbicides, including the widely used Agent Orange (named after the color of the barrel in which it came), were sprayed from airplanes, some were dispersed from tanks on the backs of soldiers. Use of the herbicides sparked considerable controversy, first from groups concerned about the use of chemical agents in warfare and then from veterans who complained about the health effects of these herbicides. In 1970, Congress directed the National Academy of Sciences to study the ecological and physiological effects of the use of herbicides in South Vietnam. In a 1974 report, an NAS committee did not find definitive evidence of an association between exposure to herbicides and ill health, although the committee's primary focus was on the environmental, not the individual, effects of herbicides. Complaints from veterans kept the issue alive in the media and in Congress and led to many epidemiological investigations, some conducted with the oversight of the Institute of Medicine, that tried to determine the link between exposure to herbicides and various cancers. Complex political, epidemiological, and legal issues were involved. Congress hoped that the Institute of Medicine could provide an authoritative version of the scientific and medical "facts" pertinent to the issue.[7]

Kenneth Shine realized the political sensitivity of the case and, for this reason, took extra precautions to ensure that the members of the study committee were not themselves participants in the medical or scientific aspects of the controversy. The very fact that the Institute of Medicine, rather than the National Research Council acting on behalf of the National Academy of Sciences, conducted the study indicated the IOM's rise in status since the original NAS study of herbicides in the early 1970s. To chair the study committee, the IOM chose Harold Fallon, the dean of the University of Alabama School of Medicine, and David Tollerud, the director of occupational and environmental medicine at the University of Pittsburgh. They and 14 others, with the active assistance of project director Michael Stoto, conducted an ambitious study in which they investigated the history of the conflict, read through the considerable volume of scientific studies and reached conclusions on their validity, and considered the need for further research.[8]

At the heart of the completed report was a table that presented the committee's conclusions about the level of correlation between particular diseases and exposure to herbicides. For example, the committee found a positive association between soft-tissue sarcoma and exposure to herbicides and no association between skin cancer and exposure to herbicides. This finding failed to settle the dispute, however, because most of the evidence the committee assembled came from studies of people exposed to dioxin or herbicides in occupational and environmental settings, rather than during the Vietnam War. In fact, the committee knew that most Vietnam veterans had lower levels of exposure to herbicides than did the subjects of the other studies, but the exact level of exposure of Vietnam veterans remained unknown. To get at this question of exposure levels, on which many previous investigators had floundered, the committee made recommendations to facilitate the necessary research, for example, that the Department of Defense and the Department of Veterans Affairs should better identify those who had served in South Vietnam in their personnel and medical records. The committee suggested that the studies necessary to determine exposure levels be conducted in the most neutral way possible, which meant that much of the work should be done by "independent, nongovernmental scientific panels." In this spirit, the committee recommended that "a nongovernmental organization with appropriate experience in historical exposure reconstruction should be commissioned to develop and test models of herbicide exposure for use in studies of Vietnam veterans."[9]

In this way, the Institute of Medicine came to play a credible role in the Agent Orange controversy. Although it did not resolve this controversy, it did summarize and analyze the data in a way that was helpful to policymakers. Armed with this report, IOM leaders could testify before Congress and add a scientific voice to what inevitably was an emotional debate that involved not just Agent Orange but the larger issue of the meaning of the Vietnam War and the debt America owed its Vietnam veterans. Congress knew that the IOM was as impartial as any organization could be. At the same time, recommendations for further research, even those produced by a thorough and dispassionate review of the evidence, inevitably introduced subtle conflicts of interest. By suggesting a nongovernmental panel to examine a particular issue, as the IOM did in the Agent Orange study, it risked the charge that it was drumming up business for itself. Such conflicts were unavoidable, because the IOM was removed from the subject under study—the defense and chemical industries—but not from the business of medical research.

If the Agent Orange study exemplified the work of the Division of Health Promotion and Disease Prevention in the public service mode, the unintended pregnancy study, another project of this division in the Shine era, illustrated its public education mode. In this project, as in the project on tobacco, the IOM showed its desire to alert the nation to a public health hazard and to suggest ways of alleviating the hazard. Published in 1995 and conducted during a period of widespread interest in health care reform, the study, chaired by Harvard psychiatrist and veteran IOM member Leon Eisenberg, made the simple point that all pregnancies should be intended at the time of conception. All pregnancies should be the result of a conscious desire on the part of men and women to have children. The committee marshaled evidence to show that most pregnancies in America were unintended. Using 1987 data, the committee demonstrated that less than half of the pregnancies in America were intended and resulted in live births; 8 percent were unwanted pregnancies that resulted in live births, and 29 percent were unwanted or mistimed pregnancies that ended in abortion. The high rate of unintended pregnancies held dire public health consequences. An unwanted pregnancy diminished the chances that the mother would seek prenatal care early in the pregnancy and raised the chances that the baby would be born with a low birthweight. An unwanted pregnancy also increased the likelihood that births would occur to women who faced special medical risks or heavy socioeconomic burdens, such as adolescents, unmarried women, and women over 40. The committee stressed that the problem affected older as well as younger women.[10]

Although the problem was easy to identify, the solutions raised profound moral questions. For one thing, they involved the matters of birth control and abortion, subjects on which religious leaders held firm convictions; for another, they touched on the question of whether sexual activity on the part of young or unmarried people should be condoned. Some believed that the solution to unwanted pregnancies was abstaining from sexual intercourse. Such a view, according to the committee, reflected a denial of recent social trends—the age of first intercourse was dropping, for example—and a misreading of social statistics—most people at risk of unintended pregnancy were "beyond adolescence and many are married." The committee took the position that although sexual activity could not be controlled, its consequences could be regulated through the proper use of birth control. Hence, the committee recommended the initiation of a major education campaign to make the public aware of the social burdens caused by unwanted pregnancies. Among the goals of this campaign were improving "knowledge about contraception and reproductive knowledge" and increasing "access to contraception." The committee recommended that "an independent, public–private consortium," formed at the national level, should lead the effort. They saw this crusade as part of a common sense public health effort, similar to efforts to reduce smoking, reduce drunken driving, or promote the use of seat belts.[11]

Contrary to contemporary trends in health care finance, the committee did not hesitate to advocate an increase in public and private spending to support the campaign. It recommended, for example, that more health insurance policies cover contraceptive services and supplies, with no copayments or other forms of cost sharing. For the medically indigent, it supported using Medicaid to cover contraceptive services for two years following childbirth, and it recommended the provision of federal, state, and local funding for "comprehensive contraceptive services, especially for those low-income women and adolescents who face major financial barriers in securing such care."[12] The funding recommendations made it clear that the report was a call for birth control, a traditional item on the liberal political agenda, but with the new rationale of protecting the well-being of children and families by reducing unintended pregnancy, a rationale unrelated to population control and only tangentially related to improving the quality of life for women. The committee realized that the prevention of unintended pregnancies carried less political stigma than did such things as family planning or a woman's right to terminate a pregnancy.

The report on unintended pregnancy showed how the Institute of Medicine maintained its independence from partisan politics yet remained engaged in contemporary issues. Although the report involved a close reading of social statistics and other forms of hard data, it also contained philosophical assumptions—for example, on the moral validity of birth control—that went largely unexamined. In this regard, the Institute of Medicine often operated from within a broad area of consensus that was defined by the accepted opinions of medical and public health experts. Certain shades of opinion, libertarian views, or the views of fundamentalist religious leaders, for example, did not enter this arena.

Science and Politics

All of the IOM divisions had to resolve the tensions between staying independent of partisan politics and remaining engaged in contemporary issues. Even the Division of Health Sciences Policy, the part of the IOM that came closest to a concern for a pure science, dealt with politically sensitive issues. For example, the division administered the Forum on Blood Safety and Availability, the Forum on Drug Development, and the Roundtable on the Role of Academic Health Centers in Clinical Research and Education. Each of these convening functions involved the discussion of matters with what might be described as a high science content; each also concerned the politics of health science policy.

Perhaps the most important report from the Division of Health Sciences Policy that appeared in the Shine era concerned microbial threats to health in the United States. Published under the title of *Emerging Infections* in 1992, the report argued that the nation could not afford to become complacent about infectious disease. "The next major infectious agent to emerge as a threat to health in the United States may, like HIV, be a pathogen that has not been previously recognized," wrote study cochairmen Joshua Lederberg, a Nobel laureate, and Robert Shope, a Yale epidemiologist. The committee believed that the threat to the nation's health posed by disease-causing microbes "will continue and may even intensify in the coming years." Furthermore, in the matter of infectious diseases, there were no national boundaries. On the contrary, "there is nowhere in the world from which we are remote and no one from whom we are disconnected." Hence, the country needed to maintain surveillance against emerging infectious diseases, and such surveillance efforts should have both domestic and international components. The committee also recommended interventions in the traditional IOM style, such as "the expansion and coordination of National Institutes of Health-supported research on the agent, host, vector, and environmental factors that lead to emergence of infectious disease."[13]

As it turned out, the report proved quite prescient. In 1994, for example, public health authorities worried about the emergence of bubonic plague in India and the reemergence of drug-resistant forms of tuberculosis. For this reason, the IOM decided to follow up on the report and to support the efforts of the Centers for Disease Control and Prevention (CDC) to receive early alerts on significant microbial threats.[14] *Emerging Infections*, therefore, showed how the best IOM projects combined science and public health policy.

For *Assessing Genetic Risks*, a project of the Division of Health Sciences Policy that was completed in 1994, the IOM assembled the nation's leading experts, including University of Washington geneticist Arno Motulsky, to anticipate another problem on the boundary between science and social policy. This one concerned genetic testing. In every state, for example, newborns received genetic screening for a condition known as phenylketonuria. By performing this test, doctors could prevent a form of mental retardation. This test met the committee's conditions that screening be limited to situations in which there was a "clear indication of benefit to the newborn," a system was in place to confirm the diagnosis, and "treatment and follow-up are available for affected newborns." Many pregnant women at risk for bearing a Down's syndrome child received a test that, among other things, revealed the sex of a baby before birth. This test raised the possibility that a couple who desired a child of a particular sex might terminate a pregnancy if the fetus was of the other sex. Such a use of fetal diagnosis, according to the committee, represented "a misuse of genetic services that is inappropriate and should be discouraged by health professionals." In general, the committee tried to distinguish useful forms of diagnosis from "eugenic goals," although the distinction often proved elusive. Eliminating the population of Down's syndrome children was, after all, an eugenic goal. Advances in science raised even more difficult conundrums, as in the possibility that conditions with adult onset, such as Alzheimer's disease or certain types of cancer, might be identified in utero or in early childhood.[15]

Since the committee anticipated that this development could lead to an explosion of genetic testing, it tried to establish some appropriate rules and guidelines. Participation in genetic screening should be voluntary, and it should be accompanied by genetic counseling. Strict quality control should be imposed on both the tests themselves and the labs "reading" the tests. In addition to individual counseling, there should be public mass education programs so that people gained an understanding of what the tests might tell them and what options were available to them after they received the results. If genetic tests became more common, physicians would need more training in how to use and interpret them, as would nearly all health professionals. In addition, the country would have to anticipate the problems of stigma that the tests would create. How would employers or health insurers react to someone known to have a genetic predisposition toward Alzheimer's disease or, worse, the high probability of getting it by a certain age? The committee suggested that legislation might be required to protect such individuals from discrimination. Oversight of policy in what promised to be a difficult area might require the creation of a special government body for this specific purpose.[16]

All in all, the project showed how the IOM concerned itself not just with science but also with its legal, social, and ethical ramifications. This was also true of studies that concerned subjects closer to the everyday conduct of research. In 1994, for example, the Division of Health Sciences Policy produced a report on *Women and Health Research* that considered the ethical and legal issues involved in including women in clinical studies. The report started with broad general principles, for example, that "volunteers for clinical studies should be offered the opportunity to participate without regard to gender, race, ethnicity, or age" and then considered some of the problems unique to women. "Risks to the reproductive system," the committee, headed by Ruth Faden, director of the Program in Law, Ethics, and Health at Johns Hopkins, and Daniel Federman, dean of medical education at Harvard, noted, "should be considered in the same manner as risks to other organ systems."[17]

The Institute's Audience

During the 1990s, the IOM segmented the audiences for its reports. Some reports, such as the one on tobacco and even the one on assessing genetic risks, were intended to convey a message to the general public. A recent report of this type concerned sexually transmitted diseases (STDs), which an IOM study committee termed a "hidden epidemic." This report came complete with a vision statement, a central recommendation that there be "an effective national system for STD prevention" in the United States, and a series of strategies for implementing the recommendations and realizing the vision.[18] When the IOM issued a report of this type, it often called for a program of public education. In the case of sexually transmitted diseases, for example, the report urged that the mass media "accept advertisements and sponsor public service messages that promote condom use." Such a strategy was consistent with recommendations that the IOM had developed to respond to the AIDS epidemic.

Other IOM reports, such as the one on women and health research, were aimed not so much at the general public as at a particular subset of the medical profession. Although the report on women and health research concerned the subject of gender equity, a subject in which there was wide public interest, its recommendations tended to focus on very specific practices, such as clinical trials sponsored by the National Institutes of Health (NIH). Another Division of Health Sciences Policy report of this period discussed the barriers to choosing clinical research as a career pathway, a topic of immediate interest only to people engaged in or contemplating such careers. Tellingly, the study originated not from an external request but rather from the Board of Health Sciences Policy itself.[19]

Whether the IOM chose the general public or the health professions as the primary audience for a particular report, it also intended that the report reach another critical audience—the community of policymakers in the field of health. Every IOM report, even those that dealt with the inner details of science or matters of individual behavior, included specific recommendations for government action. For example, although much of the report on careers in clinical education was taken up with suggestions for universities and academic medical centers, the report also urged the NIH to develop a "debt relief package" for people who wished to pursue such a career. No IOM report, even the ones intended to spark mass public health campaigns, was complete without including directions for further research. The committee that was intent on stopping the use of tobacco among young people, for example, recommended not just a public education campaign but also a "youth-centered research agenda including studies of the efficacy of policy interventions."[20]

If there were exceptions to these general rules, they tended to occur in particular IOM divisions, such as the Food and Nutrition

Board or the Division of International Health. The Food and Nutrition Board, in particular, often broke the IOM mold, not because it was innately more creative or daring but rather because its subject concerned a topic of universal interest. Although the board issued its share of scientific reports, it also published practical guides for both consumers and health professionals. *Eat for Life,* which came in as colorful and attractive a cover as any IOM report ever published, was the board's 1992 "guide to reducing your risk of chronic disease." The large-print, easy-to-read publication contained such advice as "eat five or more servings of a combination of vegetables and fruits daily." The volume did not lack for scientific integrity. The information stemmed from a large Food and Nutrition Board study on the relationship between diet and chronic disease that culminated in 1989 in the publication of *Diet and Health.* Nor did the book oversimplify. The reader encountered such words as "polysaccharide," but these words always were accompanied with careful and accessible definitions that enabled a reader to put them in the appropriate context. A different sort of Food and Nutrition Board publication, *Nutrition During Pregnancy and Lactation,* was a handbook intended for health care practitioners who worked with pregnant or lactating women. The handbook detailed simple steps and procedures that the practitioner could take to ensure that women received proper nutrition through pregnancy and the early stages of motherhood. In this way, the board sought to incorporate the findings of its studies into clinical practice.[21]

Like the Food and Nutrition Board, the Medical Follow-Up Agency (MFUA), another of the National Research Council (NRC) agencies that had been transferred to the IOM in 1988, tended to do things differently from the rest of the organization. Staff members collected data from veterans' records, worked with epidemiologists and clinicians in academic medical centers and veterans' hospitals to analyze the data, and published the results in professional journals. Hence, the end product was often a peer-reviewed journal article or a scientific paper, rather than a formal report from a study committee, such as a 1991 article on "Suicide in Twins" that appeared in *Archives in General Psychiatry* and drew on a large data base, maintained by the MFUA, of twins who served in the military. In recognition of work of this type, the MFUA became a formal division of the IOM at the beginning of January 1994, and its steering committee, the Committee on Epidemiology and Veterans Follow-Up Studies, was given the status of an IOM board.[22]

With nearly 70,000 copies sold, *Nutrition During Pregnancy and Lactation: An Implementation Guide* topped the IOM best-seller lists in the period between the Institute's founding and 1994. Other reports on the list were ones through which the IOM hoped to launch public health campaigns—the two AIDS reports, the summary of the report on preventing low birthweight, and the report on the future of public health—or the ones that contained common sense advice—a 1984 report on bereavement and the Food and Nutrition Board's *Eat for Life*. The list of best-sellers for the period between 1990 and 1994 showed just how disparate the interests of the IOM were. It included reports on discovering the brain, emerging infections, preventing disability, growing up tobacco free, and broadening the base of treatment for alcohol problems.[23] Despite the IOM's desire to describe its work through a small number of cross-cutting themes, it was difficult to discover the common thread that ran through these studies and across the IOM's divisions.

Strategic Planning and the Health Security Plan

Kenneth Shine recognized that the diversity of the IOM's work made it difficult for the organization to plan its future. As a means of determining the IOM's goals, he led the organization in strategic planning activities that resulted in a mission statement and action plan approved by the IOM Council in November 1993.[24] This plan produced useful results in the form of a decrease in the time needed to complete a study, an increase in the number of studies with quick turnaround, a greater emphasis on the dissemination of study results, and the creation of a special focus on health care quality as an IOM-wide initiative. Encouraged and aware that an increase in endowment made it possible to raise the IOM to a new level, Shine initiated a similar process in 1997 in which he and his staff consulted with a broad array of IOM members, foundation officials, association executives, and health reporters on the organization's strengths and weaknesses. Not since the early days of David Hamburg's presidency had the organization undergone such a thorough reevaluation.

As might be expected, the many people who participated in the 1997 strategic planning exercise held widely contrasting views. In free-ranging discussions, foundation officials told Shine that the IOM did best when its reports involved "circumscribed" topics aimed at "well-targeted" audiences and that the Institute might be trying to do too much.[25] Association executives said that when the focus of a report was on key issues, such as the future of public health, it produced useful results, but the Institute produced too many reports each year and, in the process, lessened "the perception of a sense of direction." The executives worried, too, that the name of the organization excluded many nonphysicians and thought that it should be broadened to include the word "health."[26] Speaking bluntly, the health reporters said that although the IOM held high-quality news conferences, the results of its reports were often predictable. Reports too often contained statements so highly qualified as to make them vague or unclear and too often included the "more research needs to be done mantra."[27] Perhaps as a result, another group told Shine, the IOM reports were not necessarily being read by those who should see them. The IOM had to work not only on reaching its target audience but also on making sure that its recommendations were implemented. During a "blue-sky dinner" IOM members brainstormed on ways to disseminate and implement IOM studies in order to have a "high impact" on health care. One person suggested that the IOM create a special category of membership for journalists and other media professionals who were in a position to get the word out about the Institute and its activities.[28]

From all of this talk and careful consideration of the IOM's performance came a revised plan that listed the organization's goals, objectives, and strategies in order to conduct its mission, which was defined quite simply as "to advance and disseminate scientific knowledge to improve human health." Although the plan stressed core values such as "the joining of scientific and humanistic values," it also contained practical suggestions for reaching the IOM's goals. To cite one example, the plan suggested that each committee chair be debriefed on completion of a project about the quality of the staff and suggestions for improvement of the conduct of future studies. More ambitious strategies to reach objectives included holding briefings for all health-related federal agencies at least every two years in order to meet the goal of providing "timely and evidence-based analysis and advice on matters related to health." The plan also contained quite specific ideas on improving the dissemination process, for example, by maintaining a "listserv" that brought news of the IOM to congressional staffers by way of e-mail and extending the basic listserv approach, developed by Roger Herdman, a former Office of Technology Assessment director and an IOM senior scholar, to foundation officials.[29]

Through these strategic planning exercises, Kenneth Shine put his stamp on the organization. Although much of his work was taken up with an internal reassessment of the IOM an its operations, he also had to contend with the changes in the external health policy environment that occurred during the presidency of Bill Clinton. During Samuel Thier's presidency, the issue of AIDS provided a focus for the IOM's work. During Kenneth Shine's presidency, the issue of health care financing reform, as articulated in President Clinton's Health Security Plan, dominated much of the IOM's business.

Responding to the Clinton health insurance initiatives posed different sorts of problems than formulating a response to the AIDS crisis during the Reagan and Bush years. In the case of AIDS, the reticence of President Reagan on the issue enabled the IOM to play a primary role. As an organization in substantial disagreement with the President, the IOM enjoyed considerable freedom to set its own agenda. In the case of health insurance, the activism of President Clinton meant that the IOM would, of necessity, play a secondary role. As an organization that sympathized with the President's objectives, the IOM was forced to follow his lead. The AIDS crisis lent itself to the sorts of expertise to which the IOM and the NAS had ready access, such as scientists who studied viruses or statisticians who studied epidemics. The recommendations that resulted from IOM studies put the organization in the familiar position of demanding that the government do more, rather than less. Influencing the Health Security Plan required expertise in, among other fields, health care economics. Although economists such as Stuart Altman, Gail Wilensky, and Rashi Fein served important functions in the IOM, they were less prominent within the organization than were physicians from academic medical centers. Any recommendations that the IOM made would have to contend with the difficult matters of controlling health care costs and facing the trade-offs among competing objectives in the field of health care policy. Compared to the AIDS issue, the issue of health care financing played less well to the IOM's and the NAS's greatest strengths.

Nonetheless, the issue, which dominated the health care policy agenda during 1993 and 1994 and was of vital concern to IOM members, required a formal response. Aware of the need to steer the IOM away from partisan politics, Kenneth Shine positioned the IOM to lend its expertise to the effort without committing the organization to a definite role in the political debate. In making this decision, Shine showed that he had absorbed the lessons of David Hamburg's presidency. Whatever the short-term benefits, it was detrimental for the IOM to be closely associated with a particular administration. In early 1993, members learned that "the IOM will not propose a plan." Kenneth Shine and the IOM Council believed that "the kinds of decisions involving trade-offs between social and economic values that must lie at the heart of any such proposal are more properly part of the political process." This did not mean the IOM would remain uninvolved. On the contrary, the IOM hoped to play a visible role "as a source of authoritative, nonpartisan advice." It also expected to use its "unique convening capacity" to create a framework for analyzing the proposal. At the 1993 annual meeting in the fall, Shine reiterated the point that the IOM stood "ready to help in any way it can to evaluate this extraordinary effort which is about to take place."[30] It was as if the Institute of Medicine would act in an oversight capacity, much as it often did for research plans of the National Institutes of Health, for example.

On the one hand, the IOM kept its distance from the process through which the Health Security Act was created. On the other hand, it offered all the advice it could and even lent direct aid, for example, by sending the Robert Wood Johnson Foundation Fellows in Health Care Policy to work with the White House Task Force on Health Care Reform. In January 1993, Shine gave Hillary Clinton and Donna Shalala an IOM white paper on health care reform that consisted of a compendium of previous IOM proposals. The Division of Health Care Services directed much of its energy toward health care reform, releasing reports early in 1993 on access to health care in America and on the frayed relationship between employment and health benefits. The IOM also formed a special committee on assessing health care reform proposals, chaired by Walter McNerney, that issued a report in April 1993, setting down the broad goals for the health care reform effort.[31]

Hillary Clinton and her task force worked independently of the IOM, and the IOM played little role in the congressional debate that took place in 1993 and 1994. By the time the IOM met in the fall of 1994, the Clinton health care bill was dead. In retrospect, Shine's decision not to have an IOM health plan proved to be astute. Nonetheless, he looked on the collapse of heath care reform with real regret. He told IOM members that the nation would suffer "by virtue of an environment in which quality will come under pressure as the so-called market attempts to control costs." It seemed all the more important for the IOM to continue a three-year project it had launched on the quality of care. The project would involve what Shine described as a "whole range of activities"—from symposia and workshops to formal reports—all in an effort to understand how quality was measured in the health care system. If the Clinton health care plan had passed, the IOM would undoubtedly have played a major advisory role on the many quality of care issues the plan would have raised. Without the Clinton health care plan, the IOM's quality initiative still served as an important reminder, as the IOM Council stated in a July 1994 white paper, that "quality of care issues should be confronted with the same vigor and sophistication as will be directed at issues of access and cost."[32] In this manner, quality of care became an important motif of Kenneth Shine's presidency.

The special initiative on health care quality, as the project was formally known, began in the spring of 1994 and became a large and elaborate undertaking. It fulfilled the desire, articulated in the 1993 strategic planning exercise, for an Institute-wide project that would last for several years and consist of activities "designed to support and advance a significant policy issue." Health care reform, the Council decided, was simply too large a topic to handle. Not only was quality of care a more manageable theme, it also built on a long history of IOM work in this area. One report, in particular, on a strategy for quality assurance in Medicare, had yielded a useful definition. "Quality of care," the report noted, "is the degree to which health services for individuals and populations increase the likelihood of desired health outcomes and are consistent with current professional knowledge."[33] Using this definition, the IOM proceeded to investigate the ways in which quality of care could be evaluated and improved. The project produced a 1997 statement, issued under the signature of the presidents of the NAS's three central organizations, on "Focusing on Quality in a Changing Health Care System"; a series of special newsletters; and a number of IOM reports. As part of the initiative, the IOM also convened a National Roundtable on Health Care Quality.[34] Unlike the subject of national health insurance, in which the Institute of Medicine was one voice among many, the IOM acquired a form of collective expertise on matters related to quality of care and spoke with considerable authority on the issue.

The National Academy of Medicine?

As the IOM headed toward its 25th anniversary, it maintained a full roster of activities. These ranged from annual meetings in which members learned about the IOM's latest activities and heard from the nation's leading policymakers and researchers, to forums in which interested parties debated particular subjects, to specialized projects in which carefully appointed committees contemplated particular issues. The members were among the nation's most distinguished physicians, scientists, social scientists, and health care researchers. At the end of 1993, for example, the IOM contained no fewer than 23 Nobel laureates.

It seemed to Kenneth Shine and many members of the Council that the IOM resembled nothing so much as a National Academy of Medicine. With the passage of 25 years, it appeared appropriate for the IOM to make a formal name change. There were a number of compelling reasons for the change. The present name invited confusion between the independent Institute of Medicine and the governmental National Institutes of Health. On an international level, the name baffled foreign authorities, who could not understand why the United States chose to call its academy an institute. The word "institute" implied an organization that undertook research, and although this described part of what the IOM did, it failed to comprehend the special nature of the IOM's activities. Finally, the name implied that the IOM was somehow a second-class member of the Academy complex, which contained an academy of science, an academy of engineering, but only an institute of medicine. In retrospect, the name reflected an accident of history. When the IOM was created, Philip Handler expected the NAS to contain a number of institutes. As matters worked themselves out, the Institute of Medicine stood alone. Shine hoped to end this historical anomaly and change the organization's name to the National Academy of Medicine.

A change of this magnitude required a long series of approvals. When Shine approached Bruce Alberts, who was Frank Press's successor as head of the National Academy of Sciences, and the NAS Council, he received an encouraging response. The IOM Council was similarly positive. A final hurdle was the approval of the full NAS membership during its annual meeting. In this meeting, opposition to the name change surfaced. Dr. Francis Moore, a distinguished surgeon from Harvard who had earlier figured in the IOM's history as an opponent of the Board on Medicine's statement on heart transplants, argued against the change on the grounds that creation of the National Academy of Medicine would make it impossible for doctors to be elected to the National Academy of Sciences. Shine did not wish to push the matter on the floor of the NAS meeting; it was something that had to be done by consensus. The proposal died.

Summing Up

Walsh McDermott, who, as much as anyone, was responsible for the creation of the Institute of Medicine, would have applauded the decision to have the organization remain an institute, rather than an academy. In his work as head of the Board on Medicine, McDermott came to feel that an academy was an organization in which self-congratulation took precedence over action. He wanted an entity that was concerned not only with who would join but also with who would work on projects of social significance. It should not be an organization that contained only medical doctors. Instead, he favored an organization that embraced people from a wide range of disciplines, who might be able to help in the consideration of key questions of public health. Issues such as access to medical care and the diffusion of medical knowledge mattered to McDermott.

In the deliberations of the Board on Medicine, Irvine Page and James Shannon provided counterpoints to McDermott. Page felt that physicians needed an organization with the dignity of an academy to represent their interests. Although he too shunned the honorary aspects of an academy and sought a group of working members, he did not share McDermott's enthusiasm for the social aspects of medicine. He preferred that the organization be limited to physicians and to issues that affected the practice of medicine and the conduct of medical research. For this reason, he was wary of an entity contained within the National Academy of Sciences. In such a setting, the interests of physicians would be subsumed by other concerns. James Shannon, the former head of the National Institutes of Health, had still a different vision for the new organization. He wanted what John Cooper of the Association of American Medical Colleges described as an "honorific society for biomedical types."[35] He favored an organization that concerned itself with research issues and did not become dominated by social concerns to the detriment of scientific matters. He saw the successor to the Board on Medicine as an entity that would become a staunch defender of NIH within the political process, encouraging NIH's mission to support basic science and deflecting it from political demands that it respond to the latest medical fad.

McDermott won the argument in the short run. What emerged from the Board on Medicine was an institute, not an academy. The initial membership included not only distinguished academic doctors but also lawyers and social scientists. When it came time for the new Institute of Medicine to consider major issues related to public health and access to medical care, however, John Hogness, the first IOM president, found it difficult to initiate large-scale projects of the sort McDermott favored. Part of the reason was that the Nixon administration did not want to cede the political action on questions such as national health insurance to an organization whose political affiliation had yet to be tested. Congress and the components of the federal bureaucracy were more sympathetic, yet they did not offer the fledgling organization projects with a large philosophical bent but rather smaller projects with limited objectives. They wanted data on matters such as the cost of educating a physician, an empirical question many steps removed from the content of public policy.

Throughout the 1970s, furthermore, both Congress and the federal bureaucracy gained a greater capacity for doing their own quantitative and analytical work. Committee staffs and bureau chiefs discovered that it took the Institute of Medicine, which had to clear its work through the National Academy of Sciences bureaucracy, a great deal of time to complete a task. Since speed was so important to a policy system in which the window for change in a given area remained open for a very short time, Congress and the federal bureaucracy often looked elsewhere. The Institute of Medicine struggled.

In the interim, before the IOM developed a large program of studies, it resembled nothing so much as an honorary organization of the sort that would have been more congenial to Irvine Page than Walsh McDermott. The highlight of the early Institute of Medicine was the annual meeting, a gathering with a great deal of intellectual content that yielded important papers on health policy but was a social occasion nonetheless. Here, members greeted one another and discussed who deserved to become part of the group. The Institute of Medicine threatened to deteriorate into a social club for the country's best physicians and medical researchers.

This never happened. Foundations began to provide the Institute of Medicine with the support necessary to hire a core staff and undertake basic activities. Hogness and the IOM Council kept the organization focused on the important issues of the day, such as President Nixon's war on cancer or the Supreme Court's decision not to interfere with a woman's right to choose abortion. Still, the IOM failed to develop a source of sustained financial support or, because of the need to undertake any project the government offered it, much of a coherent program. The fact that Hogness served only three years and his successor Donald Fredrickson less than one accentuated the problems.

David Hamburg, the IOM's third president, devoted much of his term to reorganizing the IOM into a set of divisions with distinct responsibilities. His six-division scheme, later compressed into four, served as the basis for the IOM's present structure. As the IOM became better known and better organized, it attracted more grants and research contracts. During the Hamburg years, the IOM undertook studies in health manpower, quality assurance, and the relative effectiveness of polio vaccines. Each of these topics emerged as enduring themes of the IOM's work. Hamburg also engineered an informal alliance between the IOM and the Carter administration, which led to what amounted to collaborative projects on such subjects as the international dimensions of health, mental health policy, and the behavioral aspects of health.

During the Hamburg years, the distinctive strengths of the Institute of Medicine became more apparent as well. In the case of polio vaccines, for example, the government needed an external body, with a reputation for objectivity, to judge the relative effectiveness and safety of the two types of vaccines. Because the IOM represented neither the manufacturers of vaccines nor a government agency with a vested interest in a particular outcome and because it could convene the nation's leading scientific authorities to consider the matter, it was an ideal organization to undertake the study. Some matters had to be taken from the realms of Congress and the executive branch to a setting removed from partisan electoral politics. In the field of medicine, the IOM provided such a setting.

Not only did the organization serve as an impartial judge, it also validated decisions made by others. Hence, it provided an external sounding board for those who wished to advance a particular research agenda within the federal bureaucracy, such as putting more emphasis on behavioral science within the National Institutes of Health or, for those who sought a particular policy objective, such as an increase in the number of physicians engaged in primary care. The IOM gave advocates of the agenda or objective, particularly those buried within the federal government's considerable hierarchy, far more credibility and clout than they would otherwise have had. In this capacity, the IOM came close to realizing the objectives sought by James Shannon, who had wanted the Institute to serve as an advocate for NIH. Shannon might not have approved of the IOM's desire to broaden NIH to include social and behavioral sciences, but he would certainly have applauded its support of the NIH's basic mission and its desire to insulate NIH director from the whims of politicians. Indeed, advice to NIH on its organization and on the research programs of the various institutes became a staple of the IOM's activities.

The Hamburg years demonstrated the IOM's potential weaknesses as well as its strengths. Although the embrace of the Carter administration raised the IOM's level of visibility and influence, it also undermined the stance of neutrality and objectivity on which its role in the policy process depended. When the Reagan administration began in 1981, many of its appointees considered the IOM a remnant of the past regime. It therefore became a point of honor not to include the IOM in important policy deliberations. Major changes occurred in health care financing patterns during the Reagan years, and the IOM played little or no role in them.

Between 1980 and 1985, Fred Robbins reaped what Hamburg had sown. The IOM divisions continued to produce reports on subjects of interest to a particular group or subset of the federal bureaucracy. In the role of impartial arbiter, the IOM investigated the retirement age of airline pilots. It also supplied data to Congress on federal subsidies for nursing education and on the proper regulation of nursing homes and to the Department of Defense on graduate medical education in the military. The IOM assisted CDC in the investigation of toxic shock syndrome and NIH in assessing the health effects of marijuana. Largely on its own initiative, the IOM conducted influential studies of for-profit investment in health care and the prevention of low birthweight. Despite these activities, some of which, such as the nursing home study, produced tangible results in the forms of new laws and regulations, the IOM faced financial stringencies that were, at least in part, the result of the Reagan administration's hesitancy to award large contracts to the IOM. At the same time, foundations, affected by the recession of the early 1980s, began to grow wary of the IOM, disappointed that their investments had not led to a more coherent program or a more sustained record of influence. The National Academy of Sciences, under the leadership of a new president, entertained doubts about the ways in which the Institute of Medicine combined scientific analysis and policy advocacy.

These doubts about the Institute of Medicine culminated in a major investigation of its activities, funded by the foundations that supported the IOM and conducted by the NAS, in 1984. The authors of the Sproull report collected a great deal of seemingly damning information about the IOM—its lack of purpose, its uncertain place among organizations concerned with health policy, its lack of energy in raising money and seeking new projects, and its inability to issue decisive reports in a reasonable amount of time. Early in 1985, however, the National Academy of Sciences rejected the major conclusion of the Sproull report that the IOM be converted into the National Academy of Medicine and its research functions moved to the National Research Council. The NAS granted the IOM a reprieve.

The IOM cooperated with the NAS in selecting Samuel Thier as the new IOM president at the end of 1985. Thier conducted an energetic campaign to raise a permanent endowment for the IOM. Circumstances that had conspired against Fred Robbins, such as the NAS's lack of faith in the IOM, the depressed economy, and Ronald Reagan's control of the nation's political agenda, turned around for Sam Thier. He gained the respect of Frank Press; the economy improved; Reagan's influence faded during his second term; and the foundations, having subjected the IOM to such a close critique, felt obligated to participate in its recovery. With more money came a better ability to disseminate the results of the IOM's work. The drab reports of the 1970s became the colorful and skillfully edited reports of the 1980s. More money also brought more independence and flexibility.

Thier used the money and his newly won clout within the Academy to launch a public health crusade against AIDS. Money for the IOM's AIDS activities was not initially forthcoming from foundations or the federal government. To his credit, Thier decided to proceed on his own. The move paid off handsomely. This activity, more than any other, enabled the IOM to achieve a sense of visibility within the universe of health care policy. Furthermore, the Bush administration proved a much more willing ally of the IOM than the Reagan administration. As a result, the IOM remained close to AIDS policy even when the action shifted to the Presidential Commission. In the AIDS initiatives, the IOM came the closest to realizing Walsh McDermott's vision of a socially engaged organization tackling the toughest public health problems of its era.

Ken Shine inherited an active and financially prosperous organization from Thier. The problem for him was how to maintain the sense of momentum in an era when the nation's health challenges did not present the same sorts of opportunities for the IOM that they had in the second half of the 1980s. The divisions, augmented by the addition of the Medical Follow-Up Agency and the Food and Nutrition Board, continued to pour out reports. It was Shine's job to interpret the IOM's mission so as to bring the organization's activities into focus and demonstrate that they centered on a coherent theme. As he did so, he sought to preserve the IOM's basic characteristic of being engaged, yet independent, as it brought credible data to problems in health care.

As the history described here unfolded, the IOM's followers continued to seek measures of the organization's influence. In the early years, this influence of the IOM on health policy was small; in later years, particularly during the campaign against AIDS, this influence increased. Although the AIDS activities were a significant exception, the IOM gained the most influence in situations in which it faced a well-defined problem and in which outside political actors were prepared to act on the IOM's proposed solution. Hence, the IOM achieved more with its report on nursing home regulation than it did with its report on medical malpractice.

Still, to dwell on political influence misses the point of many IOM activities. The IOM was different from a professional trade association, such as the American Medical Association, that attempted to articulate and protect the professional interests of its members, or from an organization such as the Association of American Medical Colleges, often described in the popular press as a powerful lobby on behalf of academic medical centers. The difference was in part that the IOM sought to educate the general public and health policy decisionmakers on important aspects of health science and health practice. Public education took such forms as the Food and Nutrition Board's report on diet and chronic illness and the IOM report on bereavement or an even less high-profile report, such as the one on pain and disability. The Institute's convening functions represented a more private form of education. In the IOM's Forum on Drug Development, for example, administrators in the public and private sectors learned from one another. Meetings of the IOM Council, the various divisional boards, and the many steering committees all served as educational experiences for those who participated. Throughout its history, whatever its particular fortunes at the moment and whatever the results of a particular study, the IOM excelled at bringing talented and creative people together and letting them listen to one another. Even without solid empirical evidence of the sort needed for an IOM report, one can confidently assert that such interactions improved the quality of American health care.

When Walsh McDermott, Irvine Page, and James Shannon conducted their debates in the Board on Medicine, observers tended to regard each of their outlooks as distinct. An academy of medicine was different from a socially minded group concerned with health, which was different in turn from a private entity that protected the federal research mission in health. As it turned out, each of the notions was realized in the IOM's subsequent development. The IOM became a high-prestige organization that responded to the problems of academic physicians, as Page wanted. It evolved into an ally of the National Institutes of Health as they pursued their basic research missions, which Shannon wanted. Finally, although it took some time, the Institute of Medicine ultimately became a catalyst for the nation's public health campaigns, as McDermott wanted. If the IOM began in modest circumstances, it celebrated its 25th anniversary as a major factor in the formulation of the nation's health care policy.

Notes

1. Donna E. Shalala, "25th Anniversary Keynote Address," in Institute of Medicine, *2020 Vision: Health in the 21st Century. Institute of Medicine 25th Anniversary Symposium* (Washington, D.C.: National Academy Press, 1996), p. 108.

2. Institute of Medicine, Molla S. Donaldson and Kathleen N. Lohr, eds., *Health Data in the Information Age: Use, Disclosure, and Privacy* (Washington, D.C.: National Academy Press, 1994), pp. 8, 10.

3. Institute of Medicine, Gooloo S. Wunderlich, Frank A. Sloan, and Carolyne K. Davis, eds., *Nursing Staff in Hospitals and Nursing Homes: Is It Adequate?* (Washington, D.C.: National Academy Press, 1996), pp. 3, 4, 17–18.

4. *Nursing Staff in Hospitals and Nursing Homes,* p. 9.

5. Institute of Medicine, Barbara S. Lynch and Richard J. Bonnie, eds., *Growing Up Tobacco Free: Preventing Nicotine Addiction in Children and Youths—Overview* (Washington, D.C.: National Academy Press, 1994), p. 5.

6. *Ibid.,* p. 4.

7. Institute of Medicine, *Veterans and Agent Orange: Health Effects of Herbicides Used in Vietnam* (Washington, D.C.: National Academy Press, 1994), pp. 23–73.

8. *Ibid.,* pp. v–vii.

9. *Ibid.,* pp. 6–8, 16–21.

10. Institute of Medicine, Sarah S. Brown and Leon Eisenberg, eds., *The Best Intentions: Unintended Pregnancy and the Well-Being of Children and Families* (Washington, D.C.: National Academy Press, 1995), pp. 2–5.

11. *Ibid.,* pp. 8–9.

12. *Ibid.,* p. 13.

13. Joshua Lederberg, Robert E. Shope, and Stanley C. Oaks, Jr., eds., *Emerging Infections: Microbial Threats to Health in the United States* (Washington, D.C.: National Academy Press, 1992), pp. v, 1–2, 8.

14. For the IOM's follow-up to the report, see Kenneth Shine's remarks at the 1994 annual meeting, unpublished typescript, IOM offices.

15. Institute of Medicine, *Assessing Genetic Risks: Implications for Health and Social Policy* (Washington, D.C.: National Academy Press, 1994), pp. 5–8.

16. *Ibid.,* pp. 8–28.

17. Institute of Medicine, *Women and Health Research: Ethical and Legal Issues of Including Women in Clinical Studies* (Washington, D.C.: National Academy Press, 1994), pp. 5, 15.

18. Institute of Medicine, Thomas R. Eng and William T. Butler, eds., *The Hidden Epidemic: Confronting Sexually Transmitted Diseases* (Washington, D.C.: National Academy Press, 1997), pp. 27–30.

19. Institute of Medicine, William N. Kelley and Mark A. Randolph, eds., *Careers in Clinical Research: Obstacles and Opportunities* (Washington, D.C.: National Academy Press, 1994).

20. *Careers in Clinical Research,* p. 12; *Growing Up Tobacco Free,* p. 22.

21. Institute of Medicine, Catherine E. Woteki and Paul R. Thomas, eds., *Eat for Life: The Food and Nutrition Board's Guide to Reducing Your Risk of Chronic Disease* (Washington, D.C.: National Academy Press, 1992), pp. iii, 6, 29; Institute of Medicine, *Nutrition During Pregnancy and Lactation: An Implementation Guide* (Washington, D.C.: National Academy Press, 1992).

22. Institute of Medicine, *Annual Report 1991* (Washington, D.C.: National Academy Press), p. 49.

23. I am grateful to Mike Edington for supplying me with information on these matters.

24. This account of the IOM's strategic planning activities relies on documents in the files of IOM President Kenneth Shine, some of which are not yet available to researchers. Institute of Medicine, "Action Plan, 1993," approved by IOM Council on November 15, 1993.

25. "IOM Strategic Planning Activity, Conference Call with Foundation Officials," June 25, 1997, Shine Files.

26. "IOM Strategic Planning Activity, Meeting with Association Executives," March 6, 1997, Shine Files.

27. "IOM Strategic Planning Activity, Meeting with Health Reporters," March 13, 1997, Shine Files.

28. "Summary of Discussion," Blue Sky Dinners, April 1, 1997, and April 8, 1997, Shine Files.

29. Institute of Medicine, "Revised Mission, Goals, Objectives, Strategies— 1997 Strategic Planning Effort," approved by the IOM Council October 21, 1997, Shine Files.

30. "What Is IOM Doing About Health Reform?" *IOM News,* Volume 1, Issue 2 (March/April 1993), p. 1; Kenneth Shine, remarks at 1993 annual meeting, typescript obtained from IOM offices.

31. "What Is IOM Doing About Health Reform?"

32. Kenneth Shine's remarks at 1994 annual meeting, typescript, privately held by the IOM; "Protecting and Improving the Quality of Care Should Be a Key Element in Health Reform," *IOM News,* Volume 2, Issue 4 (July/August 1994), p. 1.

33. Institute of Medicine, "Steering Committee on America's Health in Transition: The IOM Special Initiative on Protecting and Improving Quality of Health and Health Care," March 15, 1994, Shine Files.

34. Copies of reports and examples of each of these undertakings are available in a specially prepared notebook, Institute of Medicine, "Special Initiative on Health Care Quality," updated October 1997, Shine Files.

35. John Cooper, Interview with Harvey Sopolsky, Sproull Committee Files, Accession 89-013-5, IOM Records.

Index

About the author . . .

Edward Berkowitz is a professor of history at George Washington University. Before coming to George Washington in 1982, he served as the first John F. Kennedy Fellow at the University of Massachusetts at Boston and as a senior staff member of the President's Commission for a National Agenda for the Eighties. He is the author, coauthor, or editor of nine other books including a biography of Institute of Medicine member Wilbur Cohen and a history of America's disability policy; he is also the author of more than 70 articles on various aspects of social welfare policy. Berkowitz was a Robert Wood Johnson Foundation Faculty Fellow in Health Care Finance during academic year 1987–1988. A graduate of Princeton University, Berkowitz received his Master's and Doctoral degrees in American History from Northwestern University.